ASPECTS ON FUNDAMENTS AND APPLICATIONS OF CONDUCTING POLYMERS

Edited by **Artur de Jesus Motheo**

Aspects on Fundaments and Applications of Conducting Polymers
Edited by Artur de Jesus Motheo

Published by InTech

Janeza Trdine 9, 51000 Rijeka, Croatia

Publishing Process Manager Molly Kaliman

Technical Editor Teodora Smiljanic

Cover Designer InTech Design Team

Additional hard copies can be obtained from orders@intechopen.com

Aspects on Fundaments and Applications of Conducting Polymers,
Edited by Artur de Jesus Motheo
p. cm.
ISBN 978-953-51-0385-1

For EU product safety concerns:
IN TECH d.o.o., Prolaz Marije Krucifikse Kozulić 3, 51000 Rijeka, Croatia,
info@intechopen.com or visit our website at intechopen.com.

Meet the editor

Dr Artur de Jesus Motheo is currently a Full Professor of Chemistry at University of Sao Paulo, one of the most prestigious universities in South America. He obtained his PhD at the same university and became a Postdoctoral Fellow at University of California, Davis, USA, under the supervision of Professor William R. Fawcett. So far, he advised 19 master and 12 doctoral theses, and published more than 100 scientific papers. He was a National Secretary of the Ibero-American Society of Electrochemistry from 2000 to 2002 and a Director of the Electrochemistry and Electroanalytical Chemistry Division of the Brazilian Chemical Society from 2006 to 2010. Professor Motheo's recent research interests are the application of electrochemical, photo-electrochemical and sonoelectrochemical methods for the degradation of endocrine disruptors, the development of materials combined with conduction polymers for corrosion protection of iron and aluminum alloys, and the study of adsorption at electrode/solution interfaces in the absence of faradaic processes.

Contents

Preface

It's been 35 years since MacDiarmid performed chemical and electrochemical doping of polyacetylene, initiating an era of polymer materials with properties hitherto attributed to metals, the conducting polymers or, more precisely, *intrinsically conducting polymers* (ICPs). In this period there have been many advances in both fundamental aspects of these materials and in the discoveries of new applications.Given the dynamics in this area, we understand that updates on the different aspects involved are necessary to follow these developments taking place around the world. With this objective we drew upon a number of distinguished authors, to contribute to this publication.

The chapters within this book have been classified according to their characteristics in fundaments and applications. Therefore, the first two chapters are dedicated to studies about permittivity, conductivity and triplet paramagnetic centers in conducting polymers. The applied aspects were organized covering the following subjects: the use of conducting polymers to minimize metallic corrosion, different points of view on the use of conducting polymers in various sensors, the use of conducting polymers to obtain nanomaterials, and finally, a fractal view of a conducting polymer membrane.

Finally, it was a privilege to interact with the researchers who contributed to this book. I would like to take this opportunity to thank everyone for their dedication and for meeting deadlines. Special thanks must be made to Ms. Molly Kaliman, Publishing Process Manager, for the attention and commitment to this task.

Professor Artur de Jesus Motheo
Interfacial Electrochemistry Group, Department of Physical Chemistry,
Institute of Chemistry of Sao Carlos, University of São Paulo
Brazil

Part 1

Fundaments

1

Triplet Paramagnetic Centers in Conducting Polymers – Study by ESR and SQUID

A.V. Kulikov
Institute of Problems of Chemical Physics, Russian Academy of Sciences, Russia

1. Introduction

Conducting polymers, *viz.*, polyaniline, polyacetylene, polypyrrole, polythiophene, polyphenylene, and many others, are interesting due to their unusual physical properties and a possibility of their diverse practical use. Researchers pay attention mainly to studies of luminescence and conductivity and their applications in microelectronic devices, photodiodes, sensors, batteries, technological membranes, *etc*. Magnetic properties are of a special interest, being tightly related to the nature of charge carriers and to fine features of polymer structure.

The frequently observed experimental linear temperature dependence of the product of magnetic susceptibility by temperature

$$\chi T = \chi_P T + C \tag{1}$$

makes it possible to divide the susceptibility into two components: the temperature-independent part χ_P and the part obeying the Curie law $\chi = C/T$ (see, *e.g.*, data for polyaniline (Wang et al., 1992; Ranghunathan et al., 1999; Kahol et al., 2005a), polythiophene (Kahol et al., 2005a), and polypyrrole (Joo et al., 2001). The origin of these two components is usually explained within the framework of the "metallic" model, which treats doped conducting polymers (in the form of both powders and films) as highly ordered metallic domains immersed into amorphous domains. The metallic domains are associated with the temperature-independent component (the Pauli susceptibility) while defects in the amorphous domains are responsible for the Curie susceptibility.

However, some experimental facts do not obey this scheme. (i) It is natural to expect that ESR lines of defects and metallic regions are of different widths but, in most cases, ESR lines of conducting polymers exhibit no superposition of the lines with different widths. (ii) Magnetic susceptibility is observed for both doped and undoped polymers (with odd and even number of electrons per polymer units), and in some cases, the $\chi T - T$ plots are not linear, *i.e.*, the susceptibility cannot be presented as the sum of two components: the Pauli and Curie susceptibilities. For instance, some samples of undoped polyaniline possess a weak susceptibility with the nonlinear $\chi T - T$ dependence (Kahol et al., 2004a). Polyacetylene and polythiophene demonstrate unusual magnetic properties. Polyacetylene (Ikehata et al., 1980; Masui et al., 1999) has a weak susceptibility in both doped and undoped states with the nonlinear $\chi T - T$ dependence. The magnetic properties of polythiophene depend on the nature of substituents in the ring. The susceptibility of undoped

polythiophene is low and increases upon doping (Chen et al., 1986), whereas the susceptibility of undoped alkyl -substituted polythiophenes is high and decreases upon doping (Colaneri et al., 1987; Čík et al., 2005), and the $\chi T - T$ plots are nonlinear in many cases. (iii) There is no correlation between the degree of crystallinity and the value of χ_P.
The existence of the Pauli susceptibility is considered to serve as a strong argument in favor of the metallic regions. However, many data show nonmetallic character of conductivity. It is asserted (Lee et al., 2006) that metallic polyaniline was first synthesized only in 2006.
If conducting polymers are nonmetals, we should search for another interpretation of the frequently observed linear dependences $\chi T - T$. The linear dependence $\chi T - T$ for undoped polyaniline and doped polyaniline with low conductivity has earlier been explained by the model of exchange - coupled polaron pairs (Kahol et al., 2004a; Kahol et al., 1999; Ranghunathan et al., 1998). The integration of susceptibility of antiferromagnetically bound pairs over broad distribution of the exchange interaction (from 0 to a maximum, with a constant weight) was shown to give the quasi-Pauli susceptibility. The ground state of these pairs is singlet, the singlet-triplet splitting is determined by the value of exchange interaction. This model can also explain the nonlinear $\chi T - T$ dependences and requires high values of the exchange interaction, up to 1000 K. In our opinion, these values are unrealistically high. We have previously shown (Kulikov et al., 2005) that the maximum known value for the exchange interaction (~1 K) is observed for distance between polyaniline chains of ~0.6 nm. We believe that the model of exchange - coupled polaron pairs remains valid under suggestion that the singlet—triplet splitting is not caused by the exchange interaction between two isolated centers, but it is a property of a particular polymer fragment, for example, tetramer, and cannot be interpreted as a result of the interaction of isolated spins. Our quantum chemical calculations of tetramer dication showed (Kulikov et al., 2007a) that for different conformations the singlet—triplet splitting can vary from –10 to +30 kJ mol^{-1} (from – 1000 to 3000 K). The authors (Kahol et al., 2004a; Kahol et al., 1999; Ranghunathan et al., 1998) decided that their model cannot be applied to high-conducting polymers, because both ESR and measurements of the low-temperature thermal capacity give close values for the density of electron states at the Fermi level (Kahol et al., 2005a,b). This conclusion seems unreliable because of difficulties in separating the thermal capacities of lattice and electrons due to an unclear anomaly of the temperature dependence of the thermal capacity at 2 K.
To explain all features of magnetic properties of conducting polymers, we proposed the "triplet" model and confirmed it by an analysis of our and literature data obtained by ESR and SQUID (Kulikov et al., 2007b, 2008, 2010a,b, 2011). According to the "triplet" model, conducting polymers consist of fragments only in singlet or triplet state (no doublet satates) with wide distribution of the singlet-triplet splitting, and magnetic properties of conducting polymers are described by an integral of fragment magnetization over this distribution.
This Chapter is a mini-review of our papers (Kulikov et al., 2005, 2007a,b, 2008, 2010a,b, 2011). The most convincing confirmation of the "triplet" model gives an analysis of the dependence of magnetization of polymers at helium temperatures on magnetic field. Most of the field dependences are simulated by the Brillouin function with spin S≈1, whereas the widespread "metallic" model predicts S=1/2.

2. The "triplet" model of paramagnetic centers in conducting polymers

We suppose that conducting polymers consist of fragments with close angles between the planes of adjacent rings. The fragments are separated from each other by sharp changes in these angles, and there is a set of conformations of these fragments resulting in variation of the singlet—triplet splitting in a wide range (Kulikov et al., 2007a) . The authors (Misurkin et al, 1994, 1996) pioneered in concluding that chains of conducting polymers are divided by chain defects into conjugated fragments of a final length. A hypothesis about the triplet nature of paramagnetism in conducting polymers was advanced in papers (Berlin et al., 1972; Vinogradov et al., 1976). Fragmentary structure of polythiophene was proposed (see (Čík et al., 2005) and references cited therein), according to which the polymer consists of fragments with parallel adjacent rings, and the coplanar character of the rings is violated by their turns relative to each other.

In our "triplet" model, the temperature and field dependences of magnetization of conducting polymers are analyzed on the basis of the scheme of energy levels shown below.

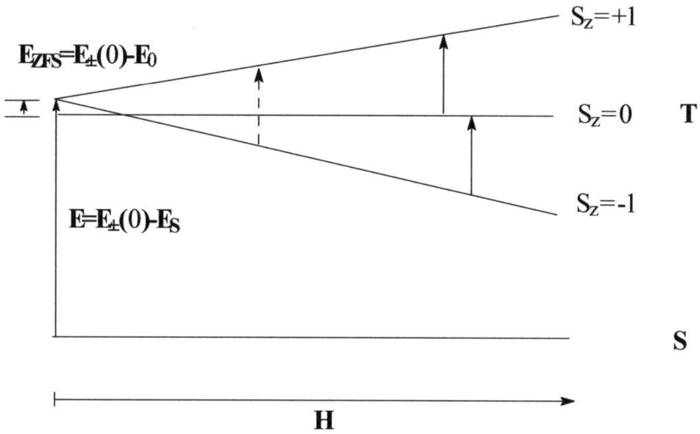

Fig. 1. Energy levels of a polymer fragment in magnetic field *H*. S and T denote singlet and triplet states, *E* is the singlet-triplet splitting, E_{ZFS} is the zero-field splitting, arrows show two allowed and one forbidden ESR transitions (Kulikov et al., 2008).

Magnetization (or magnetic moment) M of one mole of polymer elementary units is calculated by equation

$$M = \frac{g\mu_B N_A}{L}\int_{E_1}^{E_2} \frac{\exp(\frac{g\mu_B H}{kT}) - \exp(-\frac{g\mu_B H}{kT})}{\exp(\frac{g\mu_B H}{kT}) + \exp(-\frac{g\mu_B H}{kT}) + \exp(\frac{E_{ZFS}}{kT}) + \exp(\frac{E}{kT})} F(E)dE \qquad (2)$$

where g is g-factor, μ_B is the Bohr magneton, N_A is the Avogadro number, *H* is magnetic field, k is the Boltzmann constant, *F*(E) is the density of distribution of *E*, *L* is the number of polymer elementary units in polymer fragments. If $g\mu_B H/kT<<1$, $M=\chi H$, where χ is susceptibility.

Eq. (2) is easily derived on the basis of the scheme of energy levels if to take into account the Boltzmann distribution of level populations. Eq. (2) includes the length of fragments *L* (in elementary units). As a rule, the experimentally measured magnetization and susceptibility are normalized on one mole of elementary units of polymers; for instance, the unit of polyaniline holds two benzene rings. The susceptibility of fragment depends only on *E* and is independent of *L*. Therefore, with the increase in *L* the number of moles of fragments decreases and, hence, the susceptibility decreases.

The results of calculation of χT vs. *T* by Eq. (2) are shown in Fig. 2. The uniform function *F(E)*, which is constant between E_1 and E_2 and zero at other values of *E*, was used. Fig. 2

shows both linear and nonlinear curves, resembling experimental ones. At negative E_1 values, the plots are close to straight lines, and both the susceptibility components, the temperature-independent component and that obeying the Curie law, are described in the unified manner. It becomes clear why ESR lines do not reveal in most cases the superposition of two lines with different widths: both the components are of the same triplet nature. The nonlinear $\chi T - T$ dependences correspond to the case of $E_1 > 0$.

The integral in Eq. (2) can be taken in the explicit form, if the uniform (rectangular) distribution function $F(E)$ is used and $g\mu_B H/kT<<1$ and $E_{ZFS}=0$. The explicit expression of the integral facilitates simulation of experimental data; curves in Fig. 2 were calculated by this expression. Qualitatively the same dependences, obtained for the uniform distribution of the E value and presented in Fig. 2, can be obtained numerically for the Gaussian distribution of E.

Below all simulations by Eq. (2) are carried out at $E_{ZFS}=0$. For paramagnetics with S≥1 the nonzero value of E_{ZFS} results in the splitting of allowed ESR lines and arising of the weak forbidden ESR line at the half-field (see Fig. 1). The lack of the half-field forbidden transition and ESR line splitting (see Fig. 1) suggests that the zero-field splitting is less than 1 mT, or 0.01 J mol^{-1} (Kulikov et al., 2005, 2007a, 2010b).

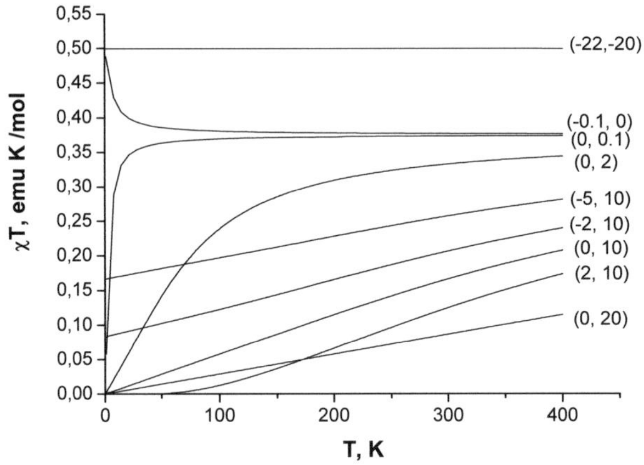

Fig. 2. Temperature dependences of the product χT calculated by Eq. (2) at $L=2$, $E_{ZFS}=0$ and the uniform function $F(E)$ with different values of E_1 and E_2 given in parentheses in kJ/mole (Kulikov et al., 2008).

3. Analysis of temperature dependences of magnetic susceptibility of conducting polymers in the framework of the "triplet" model

This part contains an analysis of our (Section 3.1) and literature (Section 3.2) temperature dependences for conducting polymers in the framework of the "triplet" model.

3.1 Effect of synthesis features, gases and heating on solutions and powders of polyaniline salts

The temperature plots of χT for polyaniline solution in m-cresol before and after heating at 423 K are presented in Fig. 3. The emeraldine base was dissolved during a month, and polyaniline transformed into the doped (protonated) form PANi(m-cresol)$_{0.5}$ (Kulikov et al., 2005). In Fig. 3, 4 and 5 the triangles oriented down, up, and sideways correspond to temperature decrease from 293 K to minimum, then to increase to 423 K, and to return to room temperature, respectively. ESR line of the solution shows no superposition of two lines and the line width is ~1 mT, which (according to (Kulikov et al., 2005)) indicates unfolded chain conformation. Characteristics of the solution remain unchanged for many months. Thus, this is a true solution of unfolded chains containing no metallic regions. Nevertheless, the linear dependence (see Fig. 3, plot *1)* is observed below room temperature, and according to Eq. (1), one could formally determine $\chi_P = 1.2 \times 10^{-4}$ emu mol^{-1} and the number of Curie spins (~0.1) per one elementary unit containing two benzene rings. The small temperature hysteresis near room temperature and the decrease in the susceptibility on heating above this temperature can be explained in the framework of the spin crossover phenomenon (Kulikov et al., 2007a). The freezing point of m-cresol is 8—10 °C.

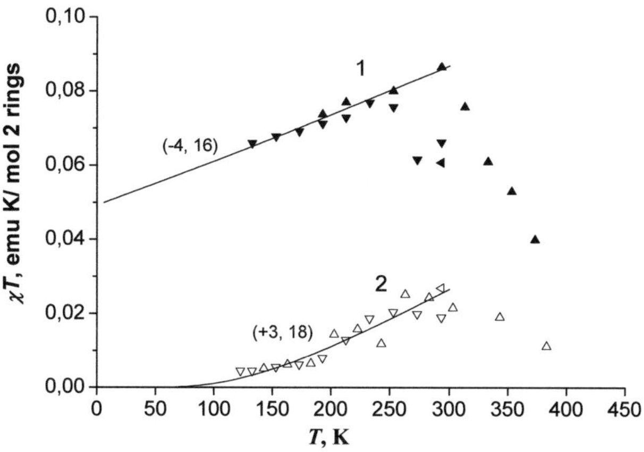

Fig. 3. Temperature dependences of χT for solution of polyaniline in m-cresol before *(1)* and after *(2)* heating at 423 K. Solid lines were calculated by Eq. (2) for $L = 4$ and the E_1 and E_2 values given in parentheses. The values of χ were measured by ESR (Kulikov et al., 2008).

After heating of the solution for 15 min at 423 K, the susceptibility decreases and the temperature dependence below room temperature becomes nonlinear (see Fig. 3, plot *2)*. After heating, the susceptibility returns at room temperature slowly (for 1 month) to the initial value (Kulikov et al., 2007a).
The nonlinear dependences cannot be explained in the framework of the "metallic" model. Plots *1* and *2* in Fig. 3 can naturally be explained in the framework of the "triplet" model. The heating changes conformations of fragments and, as a consequence, changes the distribution of the singlet—triplet splitting. Solid lines in Fig. 3 were calculated by Eq. (2) for $L = 4$ and the E_1 and E_2 values given in parentheses. The heating increases E_1 from —4 to +3 kJ mol^{-1} at an almost unchanged E_2 value (16 and 18 kJmol^{-1}).
If experimental $\chi T -\ T$ plots are nonlinear, all parameters of the "triplet" model, E_1, E_2 and L, can be determined from approximation of $\chi T -\ T$ plots by Eq. (2) (Kulikov et al., 2008). For polyaniline, L is 2 – 4; these values are close to values L=2 – 6 determined for polyaniline by the method of thermodestruction (Ivanov, 2007).
The plot $\chi T -\ T$ for powder of doped polyaniline $PANi(ClO_4)_{0.5}$ synthesized at -20° C is given in Fig. 4. The plot *in vacuo* differs from that in air. This can be explained by the change in the distribution of the singlet—triplet splitting after adsorption of dioxygen on the polymer.

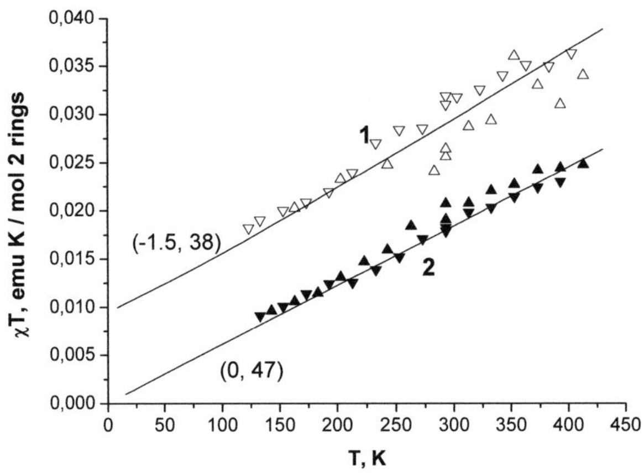

Fig. 4. Temperature dependences of the product χT for powder of doped polyaniline PANi$(ClO_4)_{0.5}$ *in vacuo* (*1*) and air (*2*). PANi was synthesized at -20° C. Solid lines were calculated by Eq. (2) for L=4 and values of E_1 and E_2 given in parentheses. The values of χ were measured by ESR (Kulikov et al., 2008).

Fig. 5 shows $\chi T - T$ plots *in vacuo* and in air for PANi$(ClO_4)_{0.5}$ synthesized at room temperature. These plots differ from those shown in Fig. 4. Thus, synthesis conditions affect the conformations of polyaniline fragments and, as a consequence, the distribution of the singlet—triplet splitting.

Plot *1* in Fig. 5 is nonlinear and cannot be simulated by Eq. (2). Plot *1* can be explained under assumption that for 3% of polymer fragments E_1 and E_2 values are negative and much lower than kT, and for remaining fragments E_1 and E_2 are 6 and 35 kJ mol^{-1}, respectively (at L = 4). In other words, we assume that the distribution of singlet-triplet splitting $F(E)$ is the sum of two rectangular functions. This kind of the distribution function was used also for simulation of $\chi T - T$ plots measured by SQUID (see below). Plot *2* in Fig. 5 measured in air is linear and can be simulated by Eq. (2) at L = 4, E_1 = 0, and E_2 = 27 kJ mol^{-1}.

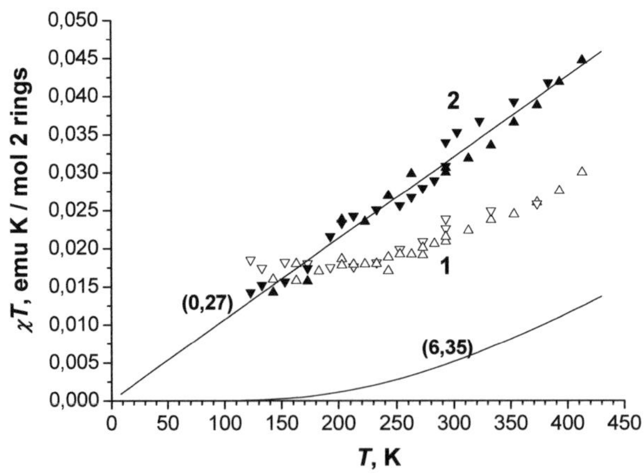

Fig. 5. Temperature dependences of the product χT for powder of PANi$(ClO_4)_{0.5}$ *in vacuo* (*1*) and in air (*2*). PANi was synthesized at room temperature. Solid lines were calculated by Eq. (2) for L = 4 and the E_1 and E_2 values indicated in parentheses. The values of χ were measured by ESR (Kulikov et al., 2008).

3.2 Literature temperature dependences for polythiophene and polyacetylene

Fig. 6 shows analysis of data (Šeršeň et al., 1996) on susceptibility of poly(3-dodecylthiophene) in the framework of our model. The authors assumed that the polymer consists of fragments, susceptibility of each fragment obeys the Curie law, but the number of fragments decreases with decreasing temperature due to recombination of fragments. They succeeded in good approximation of experimental data by formula $\chi \sim \Sigma \exp(-E_i/kT)/T$ (solid line in Fig. 6a). However, their model is not realistic because the twist of thiophene rings required for recombination of fragments is improbable in films at low temperatures. Eq. (2) with E_1=0.7 kJ/mol, E_2=8.2 kJ/mol and L=78 describes well their data (solid lines in Fig. 6b). Uncertainties in values of E_1, E_2 and L are given in Fig. 6b. Our model does not require temperature changes in chain conformation.

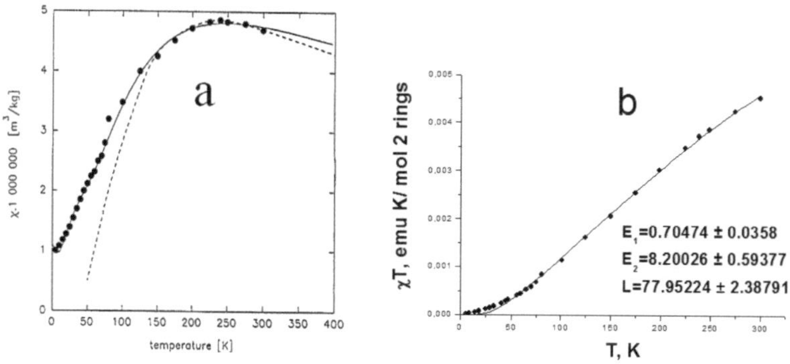

Fig. 6. Temperature dependence of magnetic susceptibility of film of poly(3-dodecylthiophene). (a) The dependence χ-T taken from (Šeršeň et al., 1996). (b) Simulation of this dependence in coordinates χT-T by Eq. (2) (Kulikov et al., 2007b). The values of E_1 and E_2 are given in kJ mol^{-1}. The values of χ were measured by a magnetometer.

Fig. 7 shows analysis of data (Masui & Ishiguro, 2001) on susceptibility of ν-trans-polyacetylene in the framework of our model. The authors explain the appreciable "spin gap" below 200 K by "spin-charge separation". Our analysis (Fig. 7b) did not reveal any phase transitions. It is worthwhile to mention that for all doping degrees except 6.6% the ESR lines are Lorentzian, without indications of superposition of lines from metallic and amorphous regions.

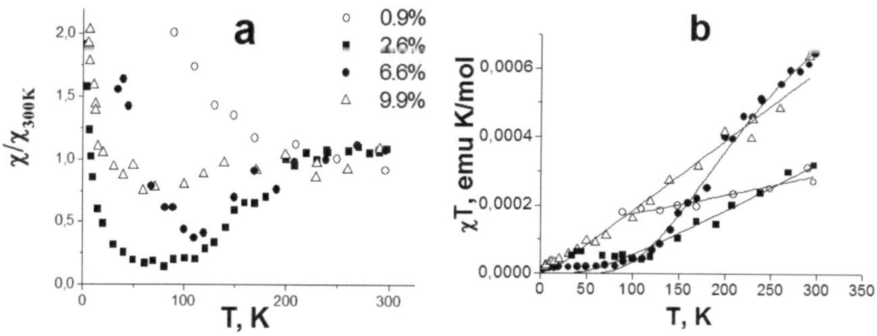

Fig. 7. Temperature dependences of magnetic susceptibility of ν-trans-polyacetylene at various degrees of doping. (a) Data taken from (Masui & Ishiguro, 2001). (b) Simulation of these data by Eq. (2) with parameters given in Table 1 (Kulikov et al., 2007b). The values of χ were measured by ESR.

Uncertainties of the parameters for polyacetylene are not given in Table 1; they are much higher than those for polythiophene because the experimental data are of bad quality and some plots in Fig. 7b are close to straight lines.

Doping, %	E_1, kJ/mol	E_2, kJ/mol	L
0.9	-2.4	145	126
2.6	1.2	87	94
6.6	4.1	4.3	540
9.9	0.2	88	64

Table 1. Parameters of Eq. (2) used for simulation of data in Fig. 7b.

4. Analysis of field dependences of magnetization of polyaniline and polypyrrole in the framework of the "triplet" model

Combined measurement of temperature and field dependences of magnetization is a severe exam for the "triplet" model. In the "metallic" model, the ratio of the temperature - independent component to the Curie component of the paramagnetic susceptibility is an experimental fact, whereas the "triplet" model provides the unified explanation for these components by Eq. (2). One can decide between the "metallic" and "triplet" models by analyzing the field dependence of magnetization of conducting polymers at low temperatures. If the "metallic" model is valid, mainly defects in amorphous domains should be observed at low temperatures because the Curie component increases at lowering temperature as 1/T, and the field dependences at helium temperature should be described by the Brillouin function (see, for instance, (Carlin et al., 1986)) with spin S= 1/2:

$$M(\eta) \sim (S+0.5)\mathrm{cth}[(S+0.5)\,\eta]-0.5\mathrm{cth}(\eta/2) \qquad (3)$$

where $\eta = g\mu_B H/kT$.

However, if the "triplet" model holds, the field dependences should be described taking into account the distribution of the singlet-triplet splitting. In this case, the field dependences may be described by the Brillouin functions with S≤1.

The $\chi T - T$ dependence for the polyaniline powder PANi(m-cresol)$_{0.5}$ is shown in Fig. 8. It is almost linear, as predicted by the "metallic" model; a slight deviation from linearity is observed at $T < 10$ K. This deviation was also reported by other authors for polyaniline and poly(ethylenedioxythiophene) (Kahol et al., 2004b, 2005a; Sitaram et al., 2005) but no explanation was given. Figure 9 demonstrates the field dependence of magnetization of polyaniline powder PANi(m-cresol)$_{0.5}$ at $T = 2$ K.

Data in Fig. 8 and 9 were corrected for the diamagnetic core by Pascal rules (see, for instance, (Carlin, 1986; Selwood, 1956)).

The temperature dependence of χT is rather well simulated by Eq. (2) (Fig. 8, solid line). To achieve a good simulation of experimental data at T<10 K, the distribution function $F(E)$ was chosen as the sum of two rectangular functions. Parameters of the distribution function were determined automatically by the Microcal Origin software.

The field dependence given in Fig. 9 is well simulated by the Brillouin functions with S=0.30 (not shown). This value of S is smaller than predicted by the "metallic" model (S =1/2). The theoretical field dependence (solid line) is similar in shape to the experimental one, and only by ~10% smaller in amplitude. Note that absolute (not relative) values of χT and M were calculated in Fig. 8 and 9 by Eq. (2).

The value S=0.3 is rather close to S=1/2 predicted by "metallic" model, and this looks not very convincing, therefore we continued our experiments and searched for field dependences in literature.

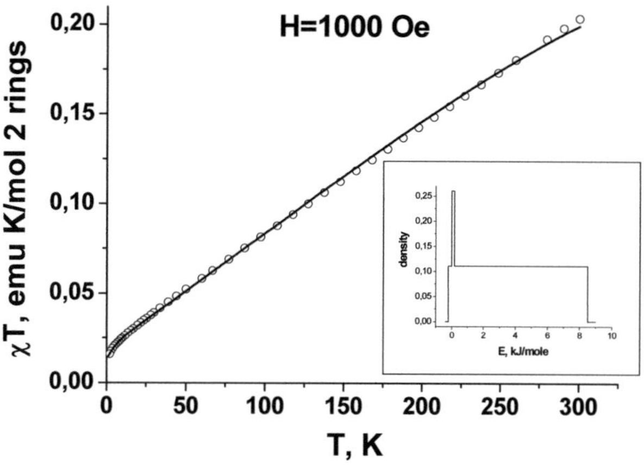

Fig. 8. Temperature dependence of χT for PANi(m-cresol)$_{0.5}$ polyaniline powder. Open circles are experimental data, solid line is the simulation of experimental data by Eq. (2) at E_{ZFS} = 0 and L = 2 for the distribution function $F(E)$ shown in the Insert. The values of χ were measured by SQUID (Kulikov et al., 2010b).

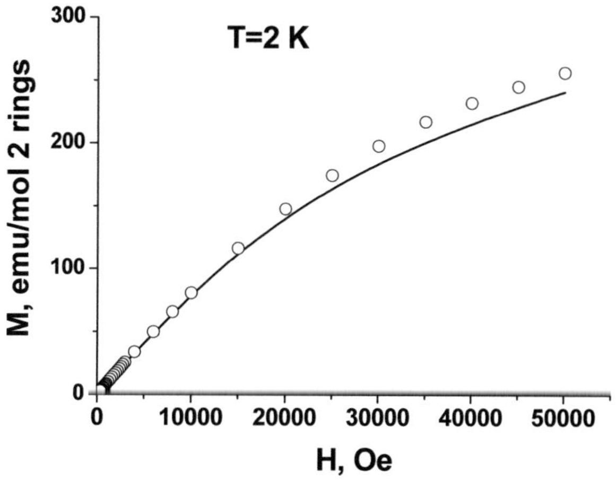

Fig. 9. The field dependence of magnetization of polyaniline powder PANi(m-cresol)$_{0.5}$ at T = 2 K. Open circles are experimental data; solid line is simulation by Eq. (2) at E_{ZFS} = 0, L = 2 and T = 2 K for the same distribution function $F(E)$, as in Fig. 8. The values of M were measured by SQUID (Kulikov et al., 2010b).

At present, we know only four field dependences of magnetization at low temperatures for conducting polymers. Fig. 10 shows two our measurements for polyaniline (including one given in Fig. 9), and two literature data for polyaniline and polypyrrole. Results of simulation of these field dependences by the Brillouin function are given in Table 2. Three field dependences are simulated by the Brillouin function with S≈1, and one our previous measurement is simulated with S=0.3. We think that this is a strong evidence in favor of the "triplet" model. In the frame of this model, the value of S is close to 1, if the share of polymer fragments with ground triplet levels (*E*<0) is high. Note that the "triplet" model can also explain the value S=0.3.

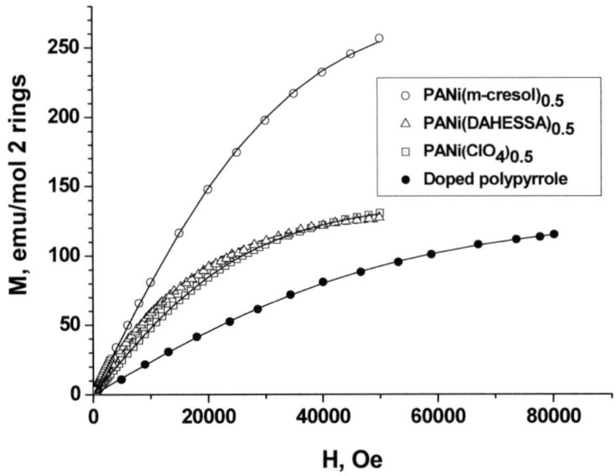

Fig. 10. Field dependences of magnetization for powders of polyaniline and polypyrrole at helium temperatures. Solid lines are simulation of the experimental results by the Brillouin function. Temperatures of measurements, values of S and references are given in Table 2. All data were obtained by SQUID. This Figure can be found in (Kulikov et al., 2010a, 2011).

Sample	Temperature, K	S	Reference
PANi(m-cresol)$_{0.5}$	2.0	0.30	Kulikov et al., 2010b
PANi(DAHESSA)$_{0.5}$	2.0	1.15	Djurado et al., 2008
PANi(ClO_4)$_{0.5}$	2.6	1.05	Kulikov et al., 2011
Doped polypyrrole	5.0	1.01	Long et al., 2006

Table 2. Parameters of the Brillouin function used for simulation of data in Fig. 10.

The authors of paper (Djurado et al., 2008) were sure that the "metallic" model is true, and simulated the field dependence for polyaniline by the Brillouin function with S=1/2, but they were forced to increase the Bohr magneton by a factor of ≈1.5. If do not make this strange increase of the universal constant, the field dependence is simulated with S=1.15. This value is a little bit higher than 1, maybe because the authors did not correct their data for the diamagnetic core. The field dependence for polypyrrole (Long et al., 2006) was not simulated by the authors.

5. Problems of the "triplet" model

Conducting polymers show no forbidden half-field ESR line and no splitting of allowed ESR lines which are typical for triplet states. Thus, for these polymers the zero-field splitting is small. Forty years ago it was explained qualitatively by the triplet state delocalization (Berlin et al., 1972). At present, the value of E_{ZFS} can be calculated by methods of quantum chemistry. We tried to calculate this splitting for doped tetramer and octamer of polyaniline by software package ORCA (Kulikov et al., 2011). It is known that magnetic properties of doped conducting polymers depend on the nature of counter-anions (Long et al., 2006), therefore we added to structures of tetramer and octamer two or four various counter-anions respectively and carried out calculations for zero net charge of these complexes. Unfortunately, the optimization procedure for these complexes without covalent bonds oligomer–counteranions was not converged.

Experimental data, for instance values of $S \approx 1$ for majority of samples, show that there is no fragments in doublet state, *i. e.,* all fragments contain even number of electrons. Maybe, the absence of fragments with odd number of electrons is due to instability of polymer structure like the Peierls instability. The Peierls' theorem states that a polymer chain with alternating spaces between adjacent elements is energetically more favorable than the chain with equal spaces. Probably, conducting polymers with even number of electrons in fragments are more stable.

SQUID and ESR are main methods of studying magnetic properties of conducting polymers. In contrast to ESR, SQUID permits to measure both temperature and field dependences of magnetization. However, magnetization measured by SQUID includes not only spin contribution described by Eq. (2) but other contributions. The "triplet" model describes only spin contribution, therefore other contributions have to be subtracted from the total magnetization. In all papers only correction for the diamagnetic core by Pascal rules is carried out. However, there is the Van Vleck paramagnetism (Van Vleck, 1932).

Both the diamagnetic and Van Vleck susceptibilities are characteristic for substances with singlet ground state and do not depend on temperature and magnetic field. These susceptibilities are of different signs and comparable absolute values, and are not detected by ESR. Paper (Kahol et al., 2004a) states that SQUID and ESR give close values of susceptibilities for one sample of polyaniline, therefore for this sample the Van Fleck susceptibility is small. However, in our work (Kulikov et al., 2011) a comparison of ESR and SQUID data for one sample of polyaniline revealed a temperature-independent contribution which is not diamagnetic one. This may be explained by appreciable the Van Vleck contribution.

The diamagnetic and Van Vleck contributions are not important at helium temperatures, because they are temperature-independent and the Curie contribution proportional to $1/T$ dominates at low temperatures.

At present, we used only two methods, ESR and SQUID, to prove the "triplet" model. Other methods are required for further proof and study of details of this model. Two methods could be used for this purpose. (i) In the "triplet" model, all variety of experimental temperature dependences of χT are explained by variety of the distribution functions $F(E)$, therefore it is important to measure this function by direct methods. Low-lying triplet levels (10 kJ/mol ~ 1000 cm^{-1}) could be detected as a low-intensive broad phosphorescence in IR region. (ii) There are other direct methods of determining the value of spin S by pulsed ESR. For instance, the spin multiplicity was confirmed by nutation spectroscopy to be $S=1/2$ for spin soliton in a π-conjugated ladder polydiacetylene (Ikoma et al., 2002). It would be interesting to compare results of study of a conducting polymer by nutation spectroscopy and SQUID (field dependence).

6. Conclusion

To explain all features of magnetic properties of conducting polymers, we proposed the "triplet" model and confirmed it by analysis of our and literature data obtained by ESR and

SQUID. According to the "triplet" model, conducting polymers consist of fragments only in singlet or triplet state (no doublet states) with wide distribution of singlet-triplet splitting, and magnetic properties of conducting polymers are described by an integral of the fragment magnetization over this distribution. The "triplet" model is alternative to the "metallic" model which is commonly accepted.

The most plain, convincing and reliable evidence in favor of the "triplet" model gives an analysis of our and literature data for polyaniline and polypyrrole. The analysis shows that the field dependences of magnetization of conducting polymers at helium temperatures are often described by the Brillouin function with $S \approx 1$, whereas the widespread "metallic" model predicts $S=1/2$. The "triplet" model describes only spin contribution, therefore other contributions have to be subtracted from the total magnetization. At helium temperatures, other contributions are insignificant.

In the "metallic" model, the ratio of the Pauli to Curie contributions of susceptibility is experimental fact and is determined by the share of metal and amorphous regions in a polymer. The "triplet" model simulates in the unified way both the temperature and field dependences; the absolute values of magnetization at various temperatures and fields are simulated rather than shapes of the dependences.

The "triplet" model is able to explain such features of temperature dependences of χT for polyaniline, polyacetylene and polythiophene, as nonlinearity of these dependences, and the effect of heating and gases on these dependences.

7. References

Berlin A.A., Vinogradov G.A., & Ovchinnikov A.A.(1972) On the nature of paramagnetism in macromolecules with conjugated bonds. *Int. J. Quant. Chem.*, Vol. 6, pp. 263-269.

Carlin R. (1986). *Magnetochemistry*, Springer-Ferlag, ISBN 3-540-15816-2, Berlin Heidelberg New York Tokio.

Chen J., Heeger A. J. & Wudl F. (1986). Confined soliton pairs (bipolarons) in polythiophene: In-situ magnetic resonance measurements. *Solid State Commun.*, Vol. 58, pp. 251-257.

Čík G., Šeršeň F. & Dlháň L. (2005). Thermally induced transitions of polarons to bipolarons in poly(3-dodecylthiophene). *Synth. Met.*, Vol. 151, pp. 124-130.

Colaneri N., Nowak M., Spiegel D., Hotta S. & Heeger A. J. (1987). Bipolarons in poly-(3-methylthiophene): Spectroscopic, magnetic and electrochemical measurements. *Phys. Rev. B*, Vol. 36, pp. 7964-7968.

Djurado D., Pron A., Travers J.-P., Duque J.C.S., Pagliuso P.G., Rettori C., Chinaglia D.L. & Walmsley L. (2008). Magnetic field dependent magnetization of a conducting plasticized poly(aniline) film. *J. Phys.: Condens. Matter*, Vol. 20, 285228, 7pp.

Ikehata S., Kaufer J., Woerner T., Pron A., Druy M. A., Sivak A., Heeger A. J. & MacDiarmid A. G. (1980). Solitons in Polyacetylene: Magnetic Susceptibility. *Phys. Rev. Lett.*, Vol. 45, pp. 1123-1126.

Ikoma T., Okada S., Nakanishi H., Akiyama K., Tero-Kubota S., Möbius K. & Weber S. (2002). Spin soliton in a p-conjugated ladder polydiacetylene. *Phys. Rev. B*, Vol. 66, 014423, 9pp.

Ivanov V. F. (2007) Structura i svoistva polianilina v interpolimernykh kompleksakh (Structure and properties of polyaniline in interpolymeric complexes), *doctoral dissertation*, Moscow, 2007 (in Russian).

Joo J., Lee J. K., Baeck J.S., Kim K. H., Oh E. J. & Epstein J. (2001). Electrical, magnetic and structural properties of chemically and electrochemically synthesized polypyrroles. *Synth. Met.*, Vol. 117, pp. 45-51.

Kahol P. K., Raghunathan A., McCormick B. J. & Epstein A.J. (1999). High temperature magnetic susceptibility studies of polyanilines. *Synth. Met.*, Vol. 101, pp. 815-816.

Kahol P. K., Raghunathan A. & McCormick B. J. (2004a). A magnetic susceptibility study of emeraldine base polyaniline. *Synth. Met.*, Vol. 140, pp. 261–267.

Kahol P. K. & Pinto N. J. (2004b). An EPR investigation of electrospun polyaniline-polyethylene oxide blends. *Synth. Met.*, Vol. 140, pp. 269-272.

Kahol P. K., Ho J. C., Chen Y. Y., Wang C. R., Neeleshwar S., Tsai C. B. & Wessling B. (2005a). On metallic characteristics in some conducting polymers. *Synth. Met.*, Vol. 151, pp. 65-72.

Kahol P. K., Ho J. C., Chen Y. Y., Wang C. R., Neeleshwar S., Tsai C. B. & Wessling B. (2005b). Heat capacity, EPR, and dc conductivity investigations of dispersed polyaniline and poly(ethylene dioxythiophene). *Synth. Met.*, Vol. 153, pp. 169-172.

Kulikov A. V., Komissarova A. S., Ryabenko A. G., Fokeeva L. S., Shunina I. G. & Belonogova O. V. (2005). Effect of chain aggregation on the conductivity and ESR spectra of polyaniline*Russ. Chem. Bull., Int. Ed.*, Vol. 54, No. 12, pp. 2794-2804.

Kulikov A. V., Komissarova A. S., Shestakov A. F. & Fokeeva L. S. (2007a). Spin crossover in polyaniline. *Russ. Chem. Bull., Int. Ed.*, Vol. 56, No. 10, 2026-2033.

Kulikov A. V., Komissarova A. S., Shestakov A. F., Shishlov M. N. & Fokeeva L. S. (2007b). On triplet nature of paramagnetic centers in conducting polymers. *Proceedings of the International Symposium "Physics and Chemistry of Processes oriented toward Development of New High Technologies, Materials, and Equipment"*, Chernogolovka, Russia, pp. 138-142, June 25-28, 2007.

Kulikov A.V., Komissarova A.S., Shishlov M.N. & Fokeeva L.S. (2008). On the triplet nature of paramagnetic centers in conducting polymers, *Russ. Chem. Bull., Int. Ed.*, Vol. 57, No. 2, pp. 324-329.

Kulikov A. V. & Shishlov M.N. (2010a). Triplet nature of paramagnetic centers in conducting polymers. Study by SQUID and ESR. *V International Conference "High-Spin Molecules and Molecular Magnets", Book of abstracts*, p. O15, N. Novgorod, Russia, September 4-8, 2010.

Kulikov A. V. & Shishlov M. N. (2010b). Nature of paramagnetic centers in polyaniline as studied by SQUID magnetometry. *Russ. Chem. Bull., Int. Ed.*, Vol. *59*, No. *5*, *pp.* 912—916.

Kulikov A.V., Shishlov M.N. & Korchagin D. V. (2011). Triplet paramagnetic centers in polyaniline. Study by SQUID and ESR. *Russ. Chem. Bull., Int. Ed.*, Vol. 60, in press.

Lee K., Cho S., Park S. H., Heeger A. J., Lee C.-W. & Lee S.-H. (2006). Metallic transport in polyaniline. *Nature*, Vol. 441, pp. 65-68.

Long Y. Chen Z., Shen J., Zhang Z., Zhang L., Xiao H., Wan M. & Duvail J. L. (2006). Magnetic Properties of Conducting Polymer Nanostructures. *J. Phys. Chem. B*, Vol. 110, pp. 23228-23233.

Masui T., Ishiguro T. & Tsukamoto J. (1999). Spin susceptibility and its relationship to structure in perchlorate doped polyacetylene in the intermediate dopant-concentration region. *Synth. Met.*, Vol. 104, pp. 179-188.

Masui T. & Ishiguro T. (2001). Spin gap behavior and electronic phase separation in doped polyacetylene. *Synth. Met.*, Vol. 117, pp. 15-19.

Misurkin I. A., Zhuravleva T. S., Geskin V. M., Gulbinas V., Pakalnis S. & Butvilos V. (1994). Electronic processes in polyaniline films photoexcited with picoseconds laser pulses: A three-dimensional model for conducting polymers. *Phys. Rev. B*, Vol. 49, pp. 7178-7192.

Misurkin I. A. (1996). Teoriya provodyashchikh polimerov (The theory of conducting polymers), *Khim. Fiz.*, Vol. 15, No. 8, pp. 110-115 (in Russian).

Raghunathan A., Kahol P. K., Ho J. C., Chen Y. Y., Yao Y. D., Lin Y. S. & Wessling B. (1998). Low-temperature heat capacities of polyaniline and polyaniline polymethylmethacrylate blends. *Phys. Rev. B*, Vol. 58, pp. 15955-15958.

Raghunathan A., Kahol P. K. & McCormic D. J. (1999). Electron localization studies of alkoxy polyanilines. *Synth. Met.*, Vol. 100, pp. 205-216.

Selwood P. W. (1956). *Magnetochemistry*, Interscience Publishers, INC., New York.

Šeršeň F., Čík G., Szabo L., Dlháň L. (1996). Role of polarons in the antiferromagnetic behaviour of poly(3-dodecylthiophene), *Synth. Met.*, Vol. 80, pp. 297-300.

Sitaram V., Sharma A., Bhat S. V., Mazoguchi K., Menon R. (2005). Electron spin resonance in doped polyaniline. *Phys. Rev. B*, Vol. 72, 035209, 7pp.

Van Vleck J. H. (1932). *The Theory of Electric and Magnetic Susceptibility*, Oxford University Press.

Vinogradov G. A., Misurkin I. A. & Ovchinnikov A. A. (1976). K voprosu o termovozbuzhdennom paramagnetizme macromolecul s sopryazhennymi svyazami (On thermoexcited paramagnetism in macromolecules with conjugated bonds).*Teoret. Eksp. Khim.*, Vol. 12, No. 6, pp. 723-730 (in Russian).

Wang Z. H., Scherr E. M., MacDiarmid A. G. & Epstein A. J. (1992). Transport and EPR studies of polyaniline: A quasi-one-dimensional conductor with three-dimensional "Metallic" states. *Phys. Rev. B*, 45, 4190-4202.

Part 2

Metallic Corrosion

2

Adhesion of Polyaniline on Metallic Surfaces

Artur de Jesus Motheo and Leandro Duarte Bisanha
University of São Paulo
Brazil

1. Introduction

Metal corrosion can cause enormous material and economic damages to general infrastructures, airplanes, reservoirs, tanks, ships, etc. The development of new materials and the association of different materials for corrosion protection have been an important area of research. In the literature, the use of conducting polymers can provide corrosion protection to metals in different environments (acid and basic aqueous media). Among the several intrinsic conducting polymers, polyaniline (PAni) stands out due to its processability, chemical stability, low cost and easy polymerization.

This chapter discusses corrosion processes and their prevention using conducting polymers, especially polyanilines, and the advantages of the use of adhesion promoters, which improve the efficiency of the coatings. The experimental results used to discuss this matter are those obtained by using iron (stainless and carbon steels) and aluminium alloys.

2. Corrosion

Corrosion is the deterioration of materials by either chemical or electrochemical action of the medium, and may or may not be associated with surface strain. When considering the use of materials in the construction of equipment or facilities, those must resist the action of the corrosive environment, as well as provide appropriate mechanical properties and manufacturing characteristics. Corrosion can be associated with different types of metallic or non-metallic materials. Considering the particular case of metallic materials, their degradation is called metallic corrosion (Fontana, 1986; Jones, 1991; Trethewey & Chamberlain, 1995).

Studies on metal corrosion are based on their importance to the increasing use of metals in all fields, specifically in the technological one. The use of large metal buildings, more susceptible to corrosion than stone structures, an increasingly aggressive environment in areas of usual applications (water, polluted air) and industrial areas (processes involving aggressive and hazardous reagents) and the use of rare and expensive metals, in some special applications (e.g., atomic energy and aerospace) are the most indicative examples of the metallic corrosion importance. Depending on the action of the corrosive medium on the material, the corrosive processes can be classified into two major groups, covering all cases of deterioration by corrosion, electrochemical corrosion and chemical corrosion. The processes of electrochemical corrosion are more common in nature and are basically characterized by their occurrence in the presence of liquid water at different temperatures with the formation of a corrosion cell in function of the movement of electrons in the metal.

Protective coatings against corrosion

Passivation is the modification of an electrode potential towards less activity (more cathodic or more noble) due to the formation of a corrosion product film, called passivating film. Some examples of metals and alloys that are passive-forming protective films are: *i)* chromium, nickel, titanium and stainless steel (passive in most corrosive media), *ii)* lead (passive in the presence of sulphuric acid), *iii)* iron (passive in the presence of concentrated nitric acid and non passive in the presence of dilute nitric acid) and *iv)* the majority of passive metals and alloys in the presence of facilities, with the exception of amphoteric metals (Al, Zn, Pb, Sn and Sb).

Besides passivating films, surfaces can be protected against corrosion by different types of protective coatings. These films are applied to metal surfaces hindering surface contact with the corrosive environment, in an attempt to minimize the degradation by the action of species on the medium. The length of protection given by a coating depends on several factors, such as type of coating (chemical nature), forces of cohesion and adhesion, thickness and permeability to the passage of electrolyte through the film.

The main mechanisms for the protection of coatings are: a barrier, anodic inhibition and cathodic protection. If protection is only a barrier as soon as the electrolyte reaches the metal surface, the corrosion process starts, whereas if there is an additional mechanism of protection (anodic inhibition or cathodic protection), the life of the coating is extended.

Different types of protective coatings can be applied to metal surfaces: *i)* anodization, which is the thickening of the protective passive layer existing in some metals, especially aluminium (the surface oxidation can be performed by either oxidants or electrochemical process and aluminium is a very common example of anodizing material), *ii)* shading, which is the reaction of the metal surface with slightly acidic solutions containing chromate (the chromate passivating layer increases the corrosion resistance of the metal surface to be protected), *iii)* phosphatization, which is the addition of a phosphate layer of the metal surface (the phosphate layer inhibits the corrosive processes and constitutes, when applied even as a thin layer, an excellent base for painting due to its roughness. The phosphating process has been widely used in the automobile and appliance industries. After the process of the metal surface degreasing, the phosphate layer is applied, followed by painting) and *iv)* organic coatings, which is the interposition of a layer of organic nature between the metal surface and the corrosive environment.

Painting is an industrial coating, usually organic, widely used for corrosion control in various types of structures and also in overhead structures, and to a lesser extent, on buried or submerged surfaces. However, there exist different types of damage leading to the deterioration of the protective film: mechanical damage caused, for example, by knocks and scrapes, damages caused by the natural action of time, such as discoloration, fading, corrosion, microcracks, etc., damages from chemical attack caused by industrial and urban pollution, damages by biological action, such as those caused by drops of resin from the trees or loose-leaf vegetation or by secretions of insects and birds.

2.1 Adhesion

When two surfaces are close to each other, an interface takes place by the action of physical and chemical forces defining an interfacial phenomenon called adhesion. The degree of attraction between the two phases defines the adhesion strength.

Considering the particular case of organic coatings, the adhesion occurs either mechanically or chemically. In the mechanical adhesion the coating penetrates the surface in its defects as pits and crevices, establishing a bond which can be improved by increasing the number of defects on the surface or its roughness. On the other hand, the chemical adhesion occurs at the metal/organic coating interface when interatomic bonds take place. These bonds can be primary (covalent or ionic bonds), secondary (dispersion forces, dipole interactions or van der Waals forces) or hydrogen bridge type. It is important to mention that metal/polymer interfacial bonds are generally secondary or hydrogen bridge type, except for epoxy resins and zinc silicates.

Adhesion problems between metallic substrates and top coats exert strong influence on the corrosion protection of a metallic surface and are caused by different factors, such as excessive film thickness and insufficient superficial cleaning. Poor adhesion allows the electrolyte to diffuse more easily into the region between the surface and the coating. The permeability of oxygen into a coating permits the occurrence of the oxygen reduction reaction, hence, an increasing OH^- concentration, which could break the metal/coating bonds. This fact leads to an increase in the polymer film detachment and further growth of blisters.

Seré et al. (Seré et al., 1996) analysed the relation between adhesion strength and corrosion resistance of carbon steel/chlorinated rubber varnish/artificial sea water systems. The authors pointed out that the adhesion of chlorinated rubber varnish onto carbon steel depends directly on the substrate surface roughness, before exposure to aggressive aqueous environments.

After immersion in aggressive environment, those samples with lower adhesion loss show a minimum corrosion level, i.e. there is a direct relationship between adhesion loss and corrosion resistance. Therefore, adhesion strength depends not only on the metal/ coating system, but also on the environment characteristics (Seré et al., 1996).

3. Polyaniline as corrosion inhibitor

Many studies have reported the use of conducting polymers as coatings (Bernard et al., 1999; Santos et al., 1998; Talman et al., 2002). Particularly, PAni has been extensively used due to its ability to protect metals against aqueous corrosion (Santos et al., 1998; DeBerry, 1985).

PAni can be obtained by either chemical or electrochemical oxidation of aniline (Trivedi, D.C., 1997). In the chemical method an oxidizing agent must be used (for instance, ammonium persulphate) and the product is obtained in powder form. In the electrochemical method, PAni is obtained in the form of ordered thin films on the electrode surface.

DeBerry (DeBerry, 1985) was the first to indicate the possibility of using PAni as a corrosion inhibitor. The author studied the electrodeposition of PAni on 410 and 430 stainless steels. He observed an anodic protection that significantly reduced the corrosion rate in sulphuric acid solution by maintaining the metal in the passive state and repassivating the damaged areas. The advantage of PAni was its effective use in acidic environments.

According to different authors (Dominis et al., 2003; Cook et al., 2004), at least three different configurations of PAni used as a corrosion protector have been reported: coatings alone, such as solution cast PAni films formed or electrochemically synthesized (Santos et al., 1998; Huerta-Vilca et al., 2003b, 2004a; Fahlman et al., 1997), coatings as primer with a conventional polymer topcoat (Dominis et al., 2003; Talo et al., 1997), and PAni blended with a conventional polymer coating or polymer coatings containing PAni as an additive (Galkowski et al., 2005; Samui et al., 2003; Sathiyanarayanan et al., 2009).

Kinlen et al. (Kinlen et al., 1997) and other authors (Spinks et al., 2002; Oliveira et al., 2009) observed that the electrochemically produced conducting polymers, as PAni, shift the corrosion potential (E_{corr}) of the metal to the passive region, maintaining a protective oxide layer on the metal and minimizing the rate of metal dissolution (Eq. 1). The reduction of oxygen to hydroxide (Eq. 2) shifts from the metal surface to the polymer/electrolyte interface and probably involves the re-oxidation of the conducting polymer (Eq. 3), stabilizing polymer coatings from cathodic disbondment. According to the above reaction sequences, the conducting polymer catalyses the oxide layer growth, protecting the metallic surface against corrosion (Oliveira et al., 2009).

$$\frac{1}{n}M + \frac{1}{m}PAni - ES^{m} + \frac{y}{n}H_2O \rightarrow \frac{1}{n}M(OH)_y^{(n-y)+} + \frac{1}{m}PAni - LE^{0} + \frac{y}{n}H^{+} \quad (1)$$

$$\frac{1}{4}mO_2 + \frac{1}{2}mH_2O + me^- \rightarrow mOH^- \quad (2)$$

$$\frac{1}{4}mO_2 + \frac{1}{2}m\,H_2O + PAni - LE^0 \rightarrow PAni - EB^{m+} + m\,OH^- \quad (3)$$

Thus, the polymer serves as a mediator between the anodic current passive layer and the reduction of oxygen in the polymer film.

4. Steels and polyanilines

Steel is the most important material in engineering followed by aluminium. Its popularity is due to the low-cost manufacturing, forming and processing and the abundance of raw materials and their mechanical properties. It can be offered in a huge number of different chemical compositions and heat treatments, microstructures, terms of conformation, geometry and surface finish. Carbon steel is steel without intentional addition of other elements, containing only carbon and four trace elements (manganese, silicon, phosphorus and sulphur) always found in steels, remaining in their composition during the manufacturing process. Stainless steels are defined as ferrous alloys that have a minimum chromium content of 11% in their constitution, because this is the element that provides corrosion resistance in certain environments.

There are many studies showing the capability of PAni acting as corrosion protection for steels in different environments. Le et al. (Le et al., 2009) deposited different PAni coatings onto 316L stainless steel, varying the cycle numbers of cyclic voltammetry (2-, 3- and 4-cycles) by electro-polymerization in 0.1 mol L^{-1} H_2SO_4 solution containing fluoride. The authors concluded that the corrosion resistance of the 316L substrate was considerably improved by the PAni coating and that the increase in the number of voltammetric cycles increased the thickness and enhanced the performance of the PAni coating due to low porosity.

Hermas et al. (Hermas et al., 2005) used electrodeposited PAni onto 304 stainless steel as protective coating in a deaerated 1.0 mol L^{-1} H_2SO_4 medium at 45°C. PAni improved the passivity of the steel, which remained passive in this aggressive medium for several weeks. After removing the PAni layer, the exposed passive oxide resisted the corrosion in acid solution for several days, in comparison to an anodically passivated film in stainless film, which broke down at once. Mrad et al. (Mrad et al. 2009) obtained PAni films on 304L stainless steel by cyclic voltammetry in different media: 0.3 mol L^{-1} $H_2C_2O_4$ or 0.3 mol L^{-1} KNO_3 (slightly basic). The authors concluded that both types of PAni coatings were able to offer a noticeable enhancement of protection against protection stainless steel corrosion process carried out in 0.5 mol L^{-1} NaCl. However, PAni electrosynthesized in oxalic acid showed the highest electroactivity, but the most porous structure properties. Also, the polymeric film synthesized in KNO_3 medium showed a better barrier property than the polymeric film obtained in oxalic acid medium.

Cook et al. (Cook et al., 2004) investigated the capacity of solution-cast PAni coating to protect mild steel in 0.1 mol L^{-1} NaCl and 0.1 mol L^{-1} HCl. The authors concluded that PAni in emeraldine base form protected mild steel in acidic and near neutral environments via inhibition of the active corrosion process rather than by anodic protection in the form of passivation.

Santos et al. (Santos et al., 1998) showed the capability of PAni in the emeraldine oxidation state to protect carbon steel and stainless steel against corrosion, in 3% NaCl aqueous solutions saturated with air. The authors related that the polymeric film is strongly adherent to the metallic substrate studied (carbon and stainless steel). The PAni film shifts corrosion potential to more positive values for both carbon steel (~100 mV) and stainless steel (~270 mV) when compared with the bare metal.

Fahlman et al. (Fahlman et al., 1997) studied the use of the emeraldine base form of PAni as a corrosion protecting undercoat on A366 cold rolled steel and iron samples when exposed to a corrosive environment consisting of a humidity chamber at 70 or 80 °C, and the performance of the PAni coatings was analyzed by X-ray photoelectron spectroscopy (XPS). The authors concluded that the mechanism for corrosion protection is anodic, the polymer passivated the metal and there was an increase in the corrosion protection when the top and interfacial oxide layers were removed prior to the polymer deposition.

Moraes and Motheo (Moraes & Motheo, 2006) deposited composites of PAni and carboxymethylcellulose (CMC) on AISI 304 surface by cyclic voltammetry using different concentrations of aniline and CMC. The authors observed that CMC interacts with PAni throughout hydrogen bonding and the morphology of the composite becomes less porous and more packed with increasing CMC concentration. The film of the composite obtained protects the surface of the steel, shifts the corrosion potential to more positive values and decreases the corrosion current. In another study, Moraes et al. (Moraes et al., 2002) synthesized PAni by chemical and electrochemical methods in phosphate buttered media. The chemically synthetized PAni was solubilized in N-methyl-pyrrrolidone and was applied on stainless steel (AISI 304). By electrochemical via, a film was deposited on the steel surface by cyclic voltammetry. The efficiency of the PAni films to inhibit the corrosion action was then studied. The authors concluded that chemically prepared PAni protects more efficiently the stainless steel in its doped state. Moraes et al. (Moraes et al., 2003a, 2003b) also showed the role of phosphate buffer solution as a moderator of the local pH variations at the polymer/electrolyte interface and the possibility of forming homogeneous and strongly adherent films. The PAni applied as coating on stainless steel acts as a corrosion inhibitor in 3% NaCl, shifting the E_{corr} to more noble values.

To illustrate the efficiency of PAni films in the corrosion protection of carbon and stainless steels, Fig. 1 depicts the potentiodynamic polarization curves for AISI 1020 carbon steel and AISI 304 stainless steel covered or not with PAni film, in NaCl 0.6 mol L^{-1}. The PAni film was formed by casting 7% (m/V) of undoped PAni solubilized in N-methyl-pyrrolidone. Corrosion studies were carried out with a potentiostat (EG&G PAR model

273A) using a saturated calomel electrode as reference, Pt plate as counter electrode and 0.5 mV s^{-1} sweep rate.

The value of E_{corr} determined for AISI 1020 carbon steel is -0.545 V. As this potential is already an oxide layer formed on the surface, with increasing potential (above the E_{corr}), this layer begins to dissolve and other processes, as pitting corrosion begin to occur. The E_{corr} of AISI 1020 coated with PAni is -0.446 V and pitting potential (E_{pit}) is -0.214 V. The protective action of PAni on the carbon steel causes the change in E_{corr} to more positive values (Fig. 1), compared to the uncoated electrode (E_{corr} = -0.545 V), due to the inhibition of redox reactions that occur at the metal/electrolyte interface. For AISI 304 stainless steel (Fig. 1), E_{corr} = -0.340 V and E_{pit}= +0.087 V. When AISI 304 is coated with PAni, there is a shift of corrosion potential to more positive values - this behaviour is similar to that of AISI 1020 carbon steel. However, for AISI 304 coated with PAni, the corrosion potential variation ($\Delta E_{corr} \cong 0.295$ V) is greater than that of the uncoated surface. This change can be justified, as described by Santos Junior et al. (Santos et al. (Santos et al., 1998)), as the attack of the species in the electrolyte, passing through the polymer layer, forming a passivation layer and interrupting the corrosion process. Fig. 2 depicts images of AISI 1020 carbon steel before and after the corrosion tests. Fig. 2a shows the risks due to the mechanical polishing and points, which can be attributed to structural defects. After the corrosion tests, the surface seem to be attacked showing (Fig. 2b) a large number of pits (corrosion points) and intergranular corrosion. The surface, which was smooth before the tests, becomes irregular and flat. The surface of AISI 1020 carbon steel, shown in Fig. 2b, is easily corroded when no corrosion protection is applied.

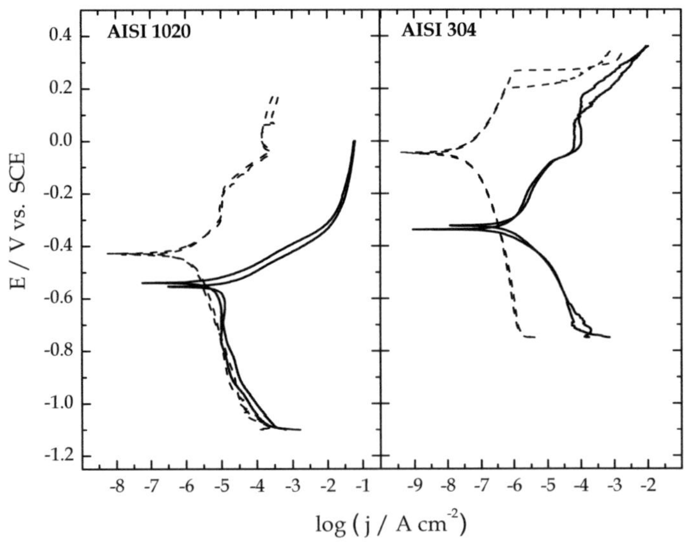

Fig. 1. Potentiodynamic polarisation curves for carbon steel AISI 1020 (left) and stainless steel AISI 304 (right) coated with PAni film (---) and bare (—) in the presence of 0.6 mol L^{-1} NaCl aqueous solution saturated with air. Sweep rate: 0.5 mV s^{-1}.

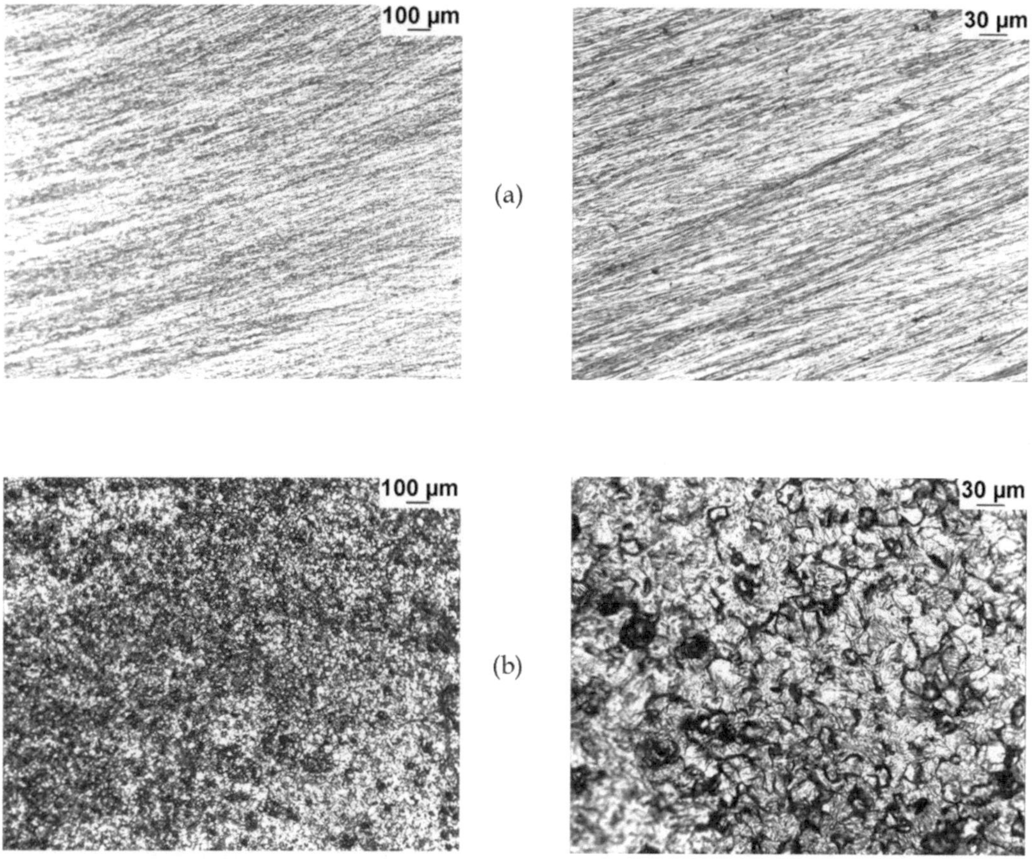

Fig. 2. Optical images of AISI 1020 carbon steel uncoated (a) before and (b) after corrosion tests. Magnification: 50x (left) and 200x (right).

Fig. 3 shows images of AISI 1020 carbon steel with the PAni film before (Fig. 3a) and after the corrosion tests (Fig. 3b) and images of the surface after removal of the PAni film (Fig. 3c). In Fig. 3a, the PAni film covering the carbon steel electrode is compact and heterogeneous.

The characteristics of the PAni film formed on the substrate depend on the characteristics of the surface of carbon steel (Fig. 3a). During the drying of the PAni film, the surface of carbon steel undergoes corrosion processes, thus forming a heterogeneous film. After the corrosion tests there is no change in the structure of the PAni film (Fig. 3b). When the PAni film is removed (Fig. 3c), it is possible to observe that the surface has been protected from corrosion, in comparison with unprotected carbon steel.

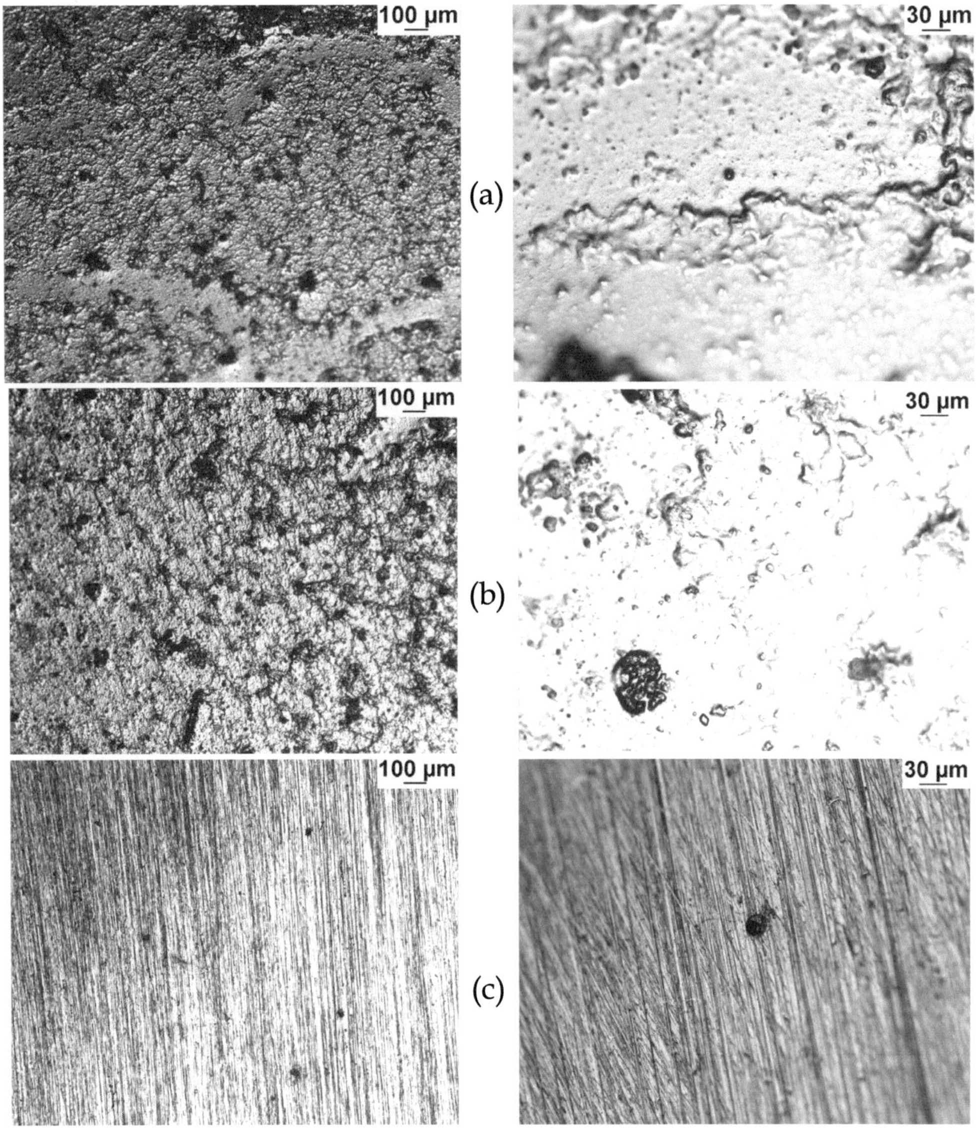

Fig. 3. Optical images of AISI 1020 carbon steel coated with PAni film (a) before and (b) after corrosion tests. (c) carbon steel surface after removal of PAni film. Magnification: 50x (left) and 200x (right).

Fig. 4 shows the images of AISI 304 stainless steel before and after the corrosion tests. The presence of risks is attributed to the mechanical polishing and the points to structural defects. After the corrosion tests one can see that the stainless steel has been attacked by chloride ions, which is evidenced by the presence of pits. Unlike what occurs in carbon steel, due to the formation of protective film of CrO_3^{-2}, for stainless steel only few pits are observed in function of the presence of chloride ions.

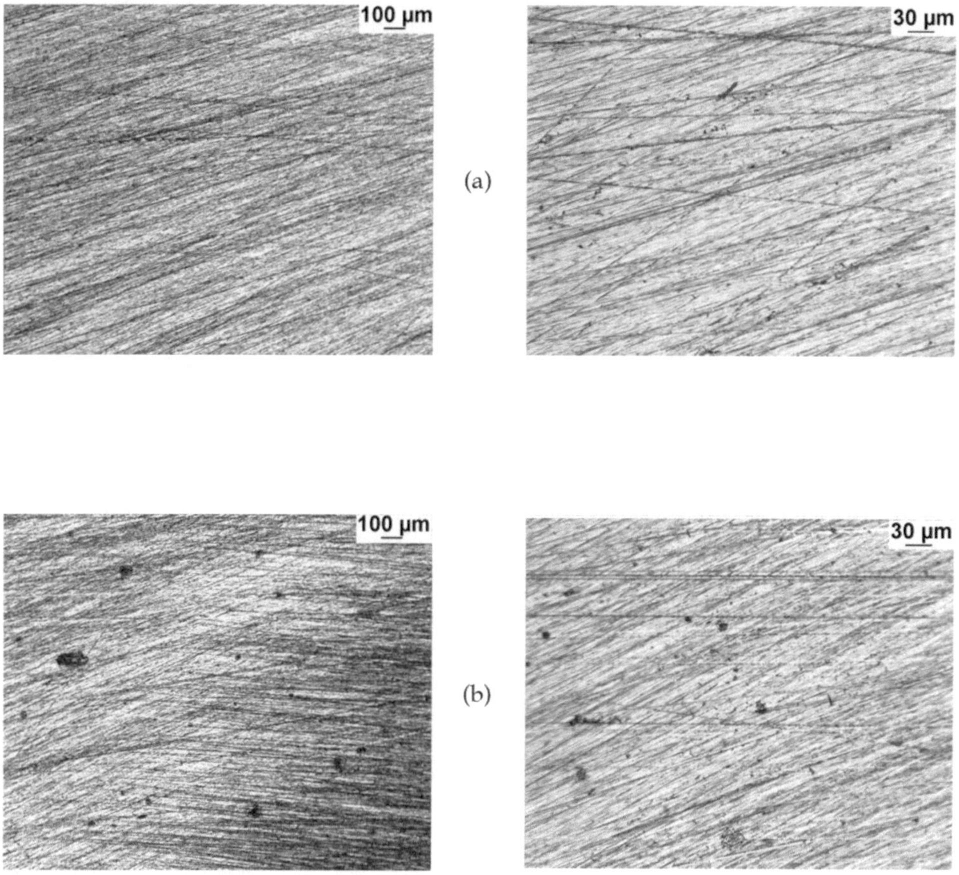

Fig. 4. Optical images of uncoated AISI 304 stainless steel (a) before and (b) after corrosion tests. Magnification: 50x (left) and 200x (right).

The PAni film formed on the surface of stainless steel is compact, uniform and homogeneous, as shown in Fig. 5a. After the corrosion tests, multiple bubbles emerged on the surface of the polymer (Fig. 5b). When the PAni film is removed (Fig. 5c), the surface does not present too many points of corrosion, evidencing the protection of the PAni film. These examples support many studies in the literature, showing the capability of PAni to protect steel surfaces, by not only barrier effect, but also anodic protection.

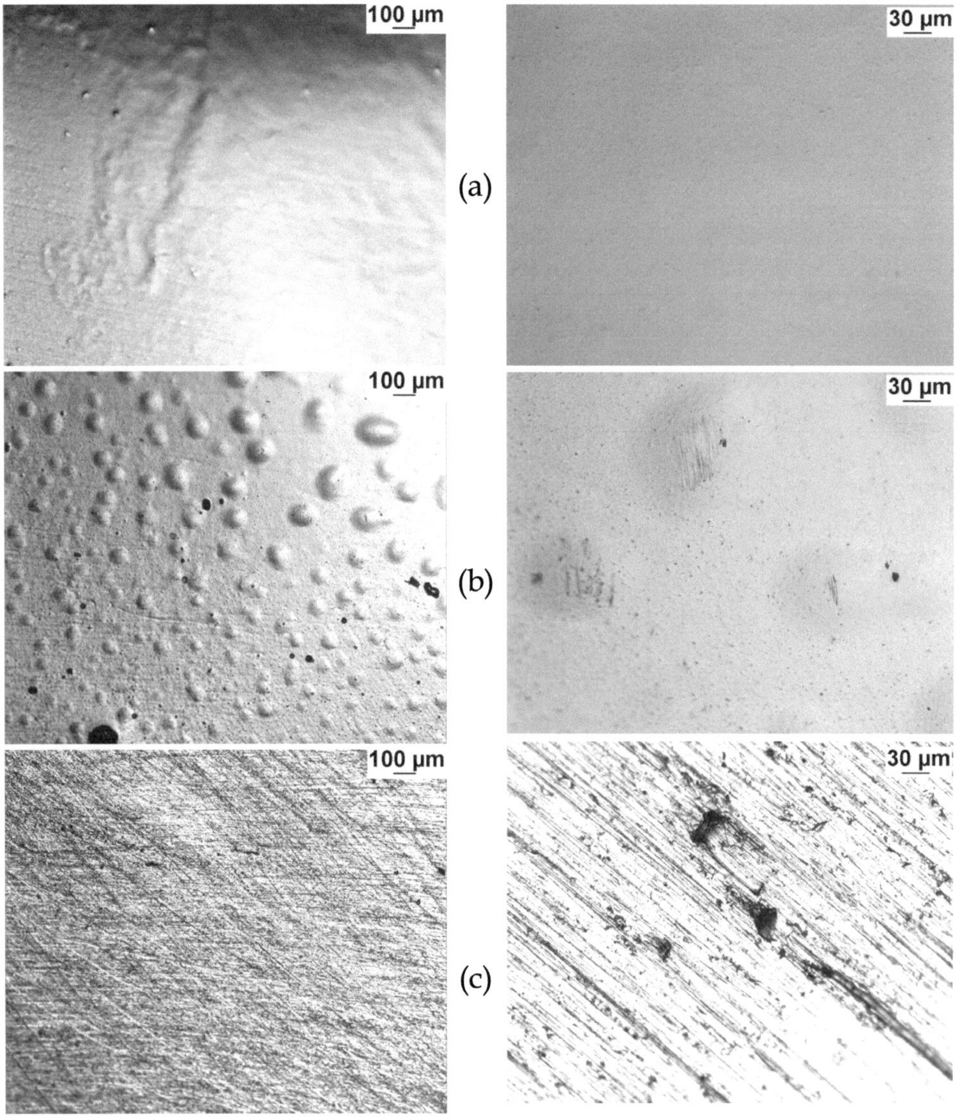

Fig. 5. Optical images of AISI 304 stainless steel coated with PAni film (a) before and (b) after corrosion tests. (c) stainless steel surface after removal of PAni film. Magnification: 50x (left) and 200x (right).

5. Aluminium, aluminium alloys and polyanilines

As presented in the latter sessions, many researchers have studied the corrosion protection of different types of steels by using chemically or electrochemically synthesized PAni, in its conductive and insulate form. Another commodity metal with increasing application is aluminium. To improve the mechanical properties, small quantities of alloying elements are added to pure aluminium, forming different aluminium alloys; however, the protection afforded is reduced by the natural passive layer.

The corrosion protection of aluminium and its alloys, especially in environments containing chloride ion, is very important, because of the increasing health risks and environmental problems with conventional coatings, like chromate and phosphate. Therefore, the coating of aluminium by PAni can become an alternative solution to the problem.

The electrodeposition of aniline on pure aluminium surface was described by Conroy and Breslin (Conroy & Breslin, 2003). The PAni films were obtained from a 1.0 mol L^{-1} tosylic acid solution containing aniline (0.132 or 0.174 mol L^{-1}) at a constant potential (1.25 V vs. SCE). After the deposition, the coated aluminium was exposed to a highly aggressive 0.5 mol L^{-1} NaCl, pH 5.85 electrolyte, using the anodic polarization technique. The PAni films shifted the E_{corr} to more positive values, with a slight increase in the corrosion resistance of aluminium. The authors related that chloride anions diffuse across the polymer and react at the underlying substrate.

Kamaraj et al. (Kamaraj et al., 2009) reported the electropolymerization of aniline on AA7075 alloy in 0.1 mol L^{-1} oxalic acid and 0.5 mol L^{-1} aniline by galvanostatatic polarization (using 20 and 15 mA for 1h). The AA7075 alloy coated with the PAni in emeraldine salt form was tested in aerated 1% NaCl by potenciodynamic polarization and impedance technique. For the film obtained using 15 mA, the corrosion current density and E_{corr} were not significantly changed in relation to the uncoated sample. For the film obtained using 20 mA, a small shift to more positive values was observed in E_{corr}. The result also confirmed the slight improvement in the corrosion protection performance. However, the authors concluded the poor corrosion resistant behaviour was due to the galvanic action of PAni and related it to the use of cerium in the post-treatment to improve the corrosion protection offered by PAni coatings.

Martins et al. (Martins et al., 2010) related the deposition of PAni films on AA6061-T6 alloy in 0.5 mol L^{-1} H_2SO_4 and 0.5 mol L^{-1} aniline using cyclic voltammetry and potentiostatic polarization. In both methods, the films obtained were adherent and presented a cauliflower structure. However, the corrosion resistance of PAni in naturally aerated 0.5 mol L^{-1} NaCl solutions showed a slight increase in the E_{corr} for coated substrate, when compared with bare metal.

The study about the capability of PAni in the emeraldine base form and self-doped sulfonated PAni form as corrosion protection of AA2024-T3 alloy was performed by Epstein et al. (Epstein et al., 1999). The results showed the efficiency of the coatings to reduce the corrosion rate when the coated alloy was exposed to 0.1 mol L^{-1} NaCl solution. The X-ray photoelectron spectroscopy analyses indicated a reduction in the copper concentration at the surface of the coated AA2024-T3 alloy samples. The authors concluded that the PAni in the different studied forms facilitated the extraction of copper from the surface of the AA2024-T3 alloy, reducing the galvanic couple between aluminium and copper, and decreasing the corrosion processes.

Fujita & Hyland (Fujita & Hyland, 2003) investigated the use of PAni (emeraldine base form) as anti-corrosion coatings for AA5005 alloy. The PAni obtained in the 1 mol L^{-1} HCl and ammonium peroxydisulfate as an oxidant was solubilized in N-methyl-pyrrolidone. This dispersion was deposited onto the alloy surfaces. The alloy substrates, uncoated and coated with PAni films, were exposed to 80% relative humidity and 30 °C for up to 4 weeks in a humidity chamber. After the corrosion tests, the samples were characterized by XPS and the authors observed that the amount of oxide grown was much smaller for coated samples than uncoated ones. However, the PAni in emeraldine base form can provide some corrosion resistance for the AA5005 alloy.

According to the classification system for aluminium alloys adopted by the Aluminum Association (Davis, 1999), zinc is the main alloying element in the AA7075 alloy (Al-Zn-Mg-Cu), while for the AA8006 alloy, the main alloying elements are iron and manganese.

The AA7075 alloy has been frequently studied because of its widespread use in the aerospace industry. On the other hand, AA8006 alloy has been successfully used for packaging and microelectronics industries, due to its high ductility and resistance to large plastic deformation in the lamination process.

Corrosion studies performed on AA7075 and AA8006 alloys uncoated and coated with PAni films were performed using potentiodynamic polarization in 0.6 mol L^{-1} NaCl aqueous solution at 0.5 mV s^{-1}. The measurements were taken with a potentiostat and a saturated calomel electrode as reference. The PAni film was formed from casting solution (undoped PAni dispersed in N-methyl-pyrrolidone).

Fig. 6 depicts the potentiodynamic polarization curves obtained for AA7075 and AA8006 alloys coated and uncoated with PAni films. The E_{corr} for AA8006 alloy was -0.913 V, while for the AA7075 alloy it was -0.768 V. These results show that the surface of the AA7075 alloy is more noble than that of the AA8006 alloy. The alloying elements present in the AA8006 (Al-Fe-Mn) provide this alloy with more susceptibility to corrosion than for the AA7075 alloy.

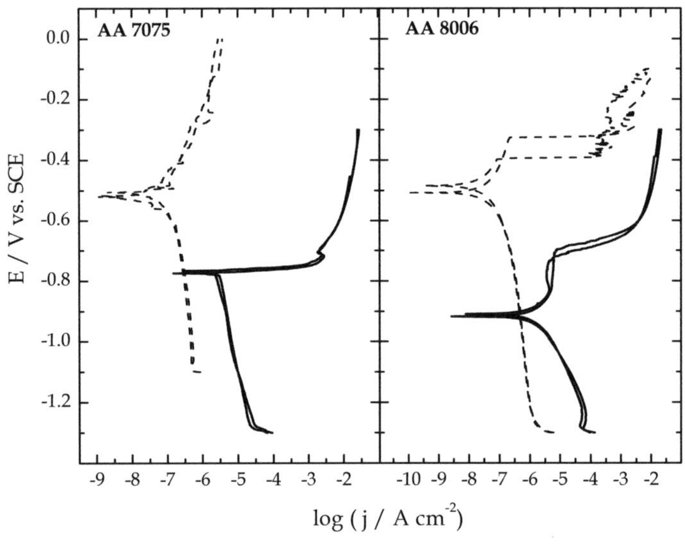

Fig. 6. Potentiodynamic polarisation curves for aluminium alloys AA7075 (left) and AA8006 (right) coated with a PAni film (---) and bare (—) in the presence of 0.6 mol L^{-1} NaCl aqueous solution saturated with air. Sweep rate: 0.5 mV s^{-1}.

For the AA8006 alloy, the pitting potential observed was -0.681 V. However, for the AA7075 alloy, E_{pit}= -0.763 V is very close to the value of E_{corr}. This result shows that for this alloy, the pitting corrosion occurs almost simultaneously with the uniform corrosion, i.e., while the passivation layer is formed, the aggressive species (Cl^- ions in the solution) attack the surface and the pitting corrosion occurs.

As observed in Fig. 6, when PAni is used as a protective coating on AA7075 and AA8006 alloys, the corrosion and pitting potentials are shifted to more positive values in relation to the uncoated electrode, providing protection anode. For the AA7075 alloy coated with PAni, E_{corr}= -0.516 V and E_{pit}= -0.,455 V, while for AA8006 alloy coated with PAni, E_{corr}= -0.501 V and E_{pit}= -0.257 V.

Fig. 7 shows optical images of the surface of AA7075 and AA8006 alloys, before and after the polarization curves. Before immersion in 0.6 mol L^{-1} NaCl solution, the surfaces of the AA7075 and AA8006 alloys were homogeneous and showed some imperfections (due to the mechanical polishing). After immersion in a corrosive solution and the polarization tests, some irregularities (holes) were observed on the surfaces, which appeared to be deep. The AA7075 alloy showed a larger number of pits in comparison with the AA8006 alloy, possibly because the values of E_{pit} and E_{corr} were very close.

The PAni film formed on the surface of the aluminium alloy is compact and uniform, covering the irregularities of the surface, as shown in Fig. 8a. After the corrosion tests, bubbles emerged on the surface of the polymer (Fig. 8b)

When the polymer film had been removed the images obtained for both alloys became similar. Fig. 8c shows the formation of a smaller number of defects on the surface coated with PAni in comparison to the uncoated one, however, the geometry of the defects was more defined.

Comparing the micrographs of the metal surface after the corrosion tests with and without coating, we can conclude that the defects are deeper when PAni is not used. On the other hand, the defects are more superficial on the surface coated with PAni.

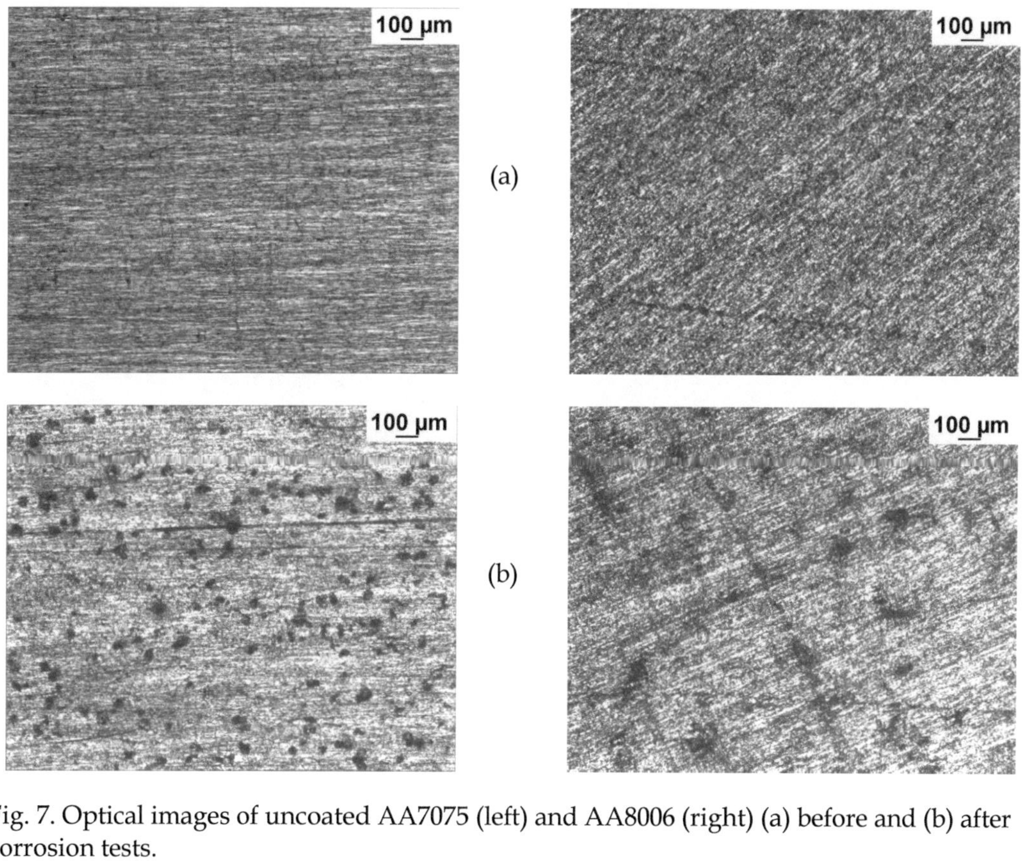

Fig. 7. Optical images of uncoated AA7075 (left) and AA8006 (right) (a) before and (b) after corrosion tests.

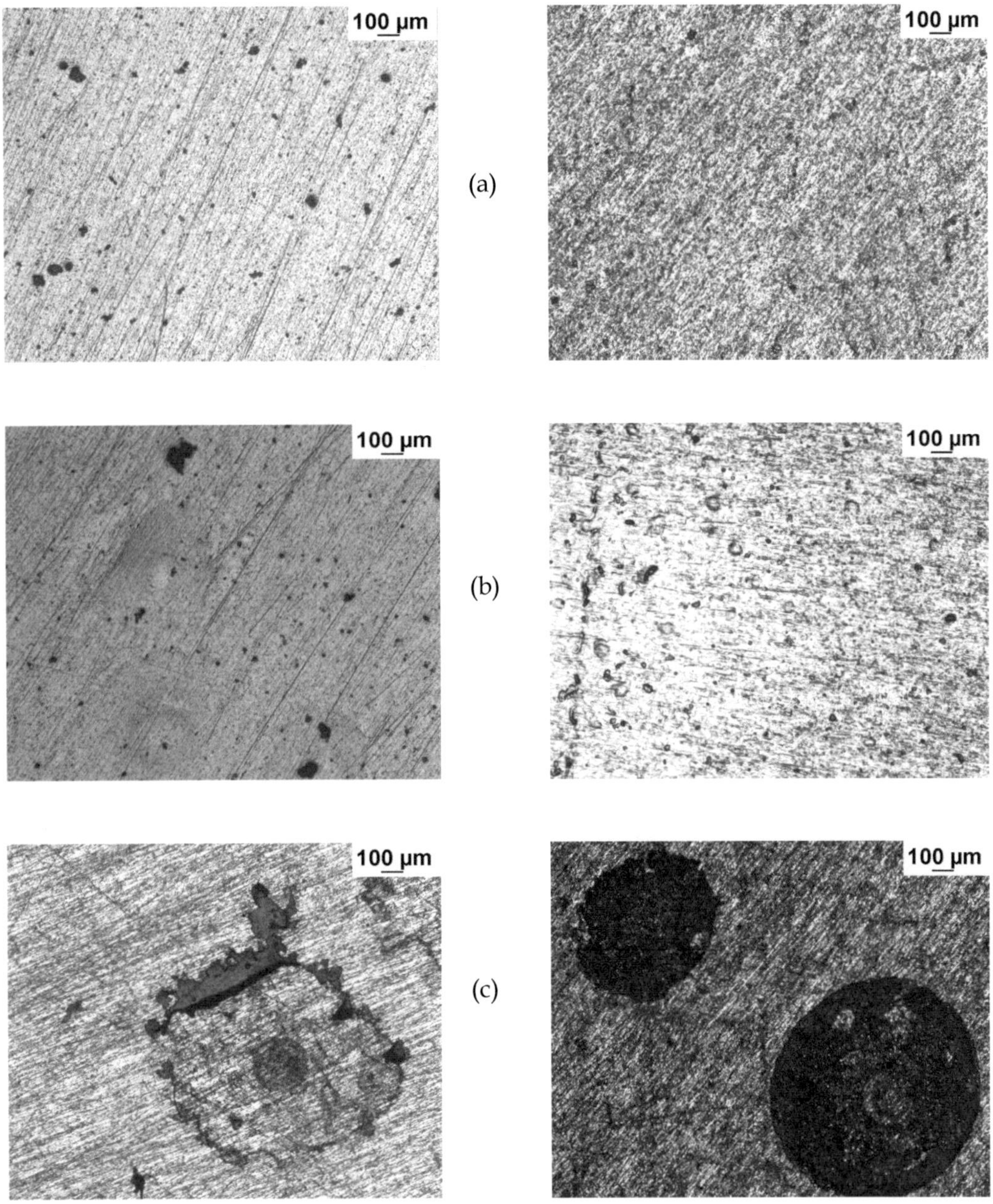

Fig. 8. Optical images of AA7075 (left) and AA8006 (right) coated with PAni films (a) before and (b) after corrosion tests. (c) aluminium alloys surface after removal of PAni film.

6. PAni adhesion on aluminium surfaces

To improve the corrosion protection of aluminium and alloys surfaces different pre-treatments, such as chromium plating, phosphating, and organic/polymeric coatings have been used. The pre-treatment of the substrates is necessary due to the natural oxides formation in passivating metals, which protect the active surface and hinder the coating adhesion. Considering that the hexavalent chromium is carcinogenic, the chromium plating or chromate conversion coatings have been abolished, and there has been a high demand for environmentally friendly surface treatments. Nevertheless, the use of organic compounds is limited because the adhesion of these materials on aluminium surfaces is very poor, requiring the use of adhesion promoters.

Among the many alternative pre-treatments, we can cite the use of SAMs (Self-assembled monolayers) and chelating agents as adhesion promoters and potential replacements for chromium-based pre-treatments. Among the SAMs, the organo-functional silanes can be employed to improve the adhesion of organic coatings to the aluminium surface. These adhesion promoters contain various functional groups, including amino, epoxy, vinyl and alkyl groups.

Bonding mechanisms of silanes to metal surfaces have been reported by different authors (Cecchetto et al., 2008; Hintze & Calle, 2006). A general mechanism can be described in the following stages: *i)* immersion of a metal into a dilute silane solution for few seconds; *ii)* spontaneous adsorption of the silanols groups (SiOH) on the surface through hydrogen bonds; *iii)* formation of covalent metallo-siloxane bonds from –SiOH groups and metal hydroxyls (HO_{Sub}). The excess of SiOH groups, adsorbed on the surface, also condenses to form a siloxane, Si–O–Si. The formed covalent bonds are assumed to be responsible for the excellent bonding of the silane film to the metal substrate.

Silva et al. (Silva et al., 2010) studied the coating combining self-assembled monolayers and PAni on AA2024 alloy. The use of octadecyltrimethoxysilane/PAni and propilmethoxysilane/PAni recovering AA2024 alloy shifted the values of corrosion and pit potentials to more positive, and the double films composed of SAM and PAni were more effective in comparison to the single SAM and PAni. The authors concluded that the double film of octadecyltrimethoxysilane/PAni presented the best corrosion protection due to the anodic protection associated with the barrier effect.

Mohseni et al. (Mohseni el al., 2006) studied the influence of amino and vinyl-silane-based treatments on the adhesion performance of an epoxy coated AA1050 alloy. The authors reported that the wettability of the AA1050 alloy increased with the use of silane-based treatment, and the surface became more hydrophilic. The adhesion strength of epoxy coated samples shows an increase in the presence of silane pre-treatment, which can be explained by the relative increase in the surface free energy of the substrates in comparison with the sample without any treatment (blank sample).

Pathak and Khanna (Pathak & Khanna, 2009) studied the combined organosilane–polyester-based waterborne coatings (SiE) using methyltrimethoxysilane, 3-glycidoxytrimethoxysilane and polyester resin for corrosion protection of AA6011 alloy. The authors observed that an increase in the concentration of organosilane in SiE coatings improved the corrosion resistance, hydrophobicity, weathering stability and hardness. The corrosion current of the SiE coated substrates decreased approximately two orders of magnitude in comparison with

the bare substrate in 3% NaCl because the coating inhibited the anodic process and acted as a barrier, blocking the contact between the electrolyte and the metal surface.
Hinze and Calle (Hinze & Calle, 2006) reported the deposition of SAMs formed from *n*-decyltriethoxysilane (DS) and *n*-octadecyltriethoxysilane (ODS) on AA2024-T3 alloy and studied their use as corrosion protection in 0.5 mol L^{-1} NaCl solution. The E_{corr} values for the modified surfaces were shifted to more positive potentials with respect to the bare surface, indicating a slight corrosion protection. However, the authors also observed that organosilanes form SAMs with a large number of defects on AA2024-T3 alloy, which probably occur over copper enriched particles found on the alloy surface.
Cecchetto et al. (Cecchetto et al., 2008) used 3-aminopropyl-triethoxysilane (APS) as primer to promote the adhesion between the AA5182 alloy substrate and PAni (in emeraldine base) coating in order to improve the corrosion protection. Potentiodynamic studies using samples coated with APS and PAni showed a reduction in the corrosion current when compared with the bare surface. The authors concluded that APS improves the adhesion of PAni coating on the alloy surface and that the corrosion resistance of PAni-coated AA5182 substrate is improved when the surface is pre-treated with an APS.
Other promoters of adhesion are the chelating agents. They have been used in the pre-treatment of metallic surfaces in the electroplating industry in order to produce decorative aluminium surfaces. Such agents can be hydrophilic or hydrophobic and can lead to the passivation of the metal by a salt or a simple formation of a blocking compound (Huerta-Vilca et al., 2003a, 2005). These authors studied the alizarin as a chelating agent in electrodeposition of PAni on aluminium surfaces (Huerta-Vilca et al., 2003a). The use of this chelating agent suppressed the evolution of hydrogen and allowed the establishment of anchor points for the growth of the polymer film. The authors also concluded that under these conditions, it is possible to electropolymerize aniline on aluminium alloys surfaces using lower aniline concentrations.

7. Anodizing

As described by Talboat and Talboat (Talboat & Talboat, 1998), anodizing is the formation of thick oxide films on metal substrates, driven by an anodic potential applied to the metal in a suitable electrolyte. The process involves the immersion of the clean metal in an acid solution, as anode, with a potential value that can be constant and appropriated or gradually increased to a maximum value. It leads to superficial films with properties that vary from a soft and porous deposit, which can be sealed to improve the corrosion resistance, to a hard one, when prepared at room temperature. Aluminum is the most commonly anodized metal and the electrolyte used could be sulphuric, phosphoric, chromic, oxalic, or boric acid (Droffelaar & Atkinson, 1995).
The morphology of the oxide formed is controlled by the nature of the electrolyte and anodizing conditions used. For example, Coz et al. (Coz et al., 2010) studied the preparation of anodic films on aluminium substrates from phosphoric acid solution. The authors were able to obtain self-supported films (without the aluminium substrate) with thickness reaching up to 130 mm, amorphous characteristics and partial hydration. The composition was determined as Al_2O_3, $0.186AlPO_4 \cdot 0.005H_2O$ and unchanged structure to temperature values up to 900 ^{0}C.

Anodizing is one of the most important processes for the corrosion protection of aluminum. Two types of anodic films can be produced: barrier and porous films. According to Mert et al. (Mert et al., 2011), the first can be formed in boric acid, ammonium borate, and ammonium tetra borate in ethylene glycol (5< pH < 7). The authors studied the synthesis of compact oxide films on aluminium, which can act as good corrosion protection, by anodizing process. The anodizing of aluminium was performed by electrolysis in 0.4 mol L^{-1} H_2SO_4 + 0.145 mol L^{-1} H_3BO_3 solution at 20ºC and the corrosion of non-anodized and anodized aluminium was investigated in 3.5% NaCl solution. The authors concluded that the ideal anodizing potential was 15 V and the convenient anodizing time was 30 min. The anodized samples (obtained in 15, 30, 60, and 120 min) shifted the E_{corr} to more positive values in relation to non-anodized aluminium. However the oxide film obtained in 30 min of anodizing showed the largest displacements of E_{corr}.
The second type is porous films and can be obtained in, for instance, sulphuric, phosphoric, chromic and oxalic acids. If the anodic oxide is slightly soluble in the electrolyte, then porous oxides are formed. This porous coating may also be coloured using organic dyes, pigment impregnation, or electrolytic deposition of various metals into the pores of the coating. After coloration, the coating is sealed. This coloured anodized aluminium combines decorative purposes with corrosion resistance, and can be used in many applications.
Based on the knowledge about porous films, Huerta-Vilca et al. (Huerta-Vilca et al., 2004b) showed a method to improve the adherence of PAni on aluminium alloys. After a galvanostatic activation in nitric acid, PAni was electrodeposited from 0.4 mol L^{-1} aniline in 0.5 mol L^{-1} H_2SO_4 by cyclic voltammetry. In the galvanostatic pre-treatment the surface was covered by a thick oxide layer with homogenously distributed pores. When PAni was electrochemically synthesized, the deposition occurred in these pores having the substrate as an anchor point. This anchorage permitted an expressive improvement in the adherence and the formation of closed polymer films covering the electrode surface with consequent improvement in the corrosion resistance. The substrate (AA1050) coated with PAni obtained by this method presented greater corrosion resistance than the samples with films growth without pre-treatment.
Another paper that illustrated the polymerization of aniline into the anodic alumina films and better corrosion protection properties was presented by Zubillaga et al. (Zubillaga et al., 2009). These authors reported the syntheses of anodic alumina films containing PAni and TiO_2 or ZrO_2 nanoparticles on AA2024-T3 alloy. The authors showed that the PAni coatings with TiO_2 nanoparticles improved the corrosion protection of the AA2024-T3 alloy in 0.005 mol L^{-1} NaCl and 0.1 mol L^{-1} Na_2SO_4 solution.

8. Conclusions

PAni has been extensively studied as a corrosion inhibitor in both doped and undoped forms on steels, aluminium and aluminium alloys surfaces. It acts as a barrier against corrosion and also anodic protection, maintaining the metal in a passive potential region. For steels, the literature has reported the good adherence of the PAni films obtained either electrochemically or by casting solution. However, to improve the adherence between PAni film and aluminium and alloys surfaces it is important to use adhesion promoters, such as SAMs and chelating agents. The adherence can also be promoted by the anodization of the surface under specific conditions. This method is usually applied in the industry to protect surfaces as well as to colour them. There still exist many possibilities to be investigated using PAni as a component in corrosion protection systems, such as blends and composites. Concerning composites, it is important to mention some preliminary researches using natural resins/PAni composites for corrosion protection, in order to improve the adherence of the film and its plasticity. The adherence of deposits as well as its influence on the corrosion protection are well known in the electroplating industry. Regarding conducting polymers, the problem is the same and the development of paints containing PAni with good adherence is already a reality. However, a new generation of coatings, called smart coatings, superior to those involving conducting polymers is in evidence. Many advantages will be attributed to them, but the adherence will be always a matter to be analysed.

9. Acknowledgments

The authors would like to acknowledge the Fundação de Amparo à Pesquisa do Estado de São Paulo (FAPESP) and the Conselho Nacional de Desenvolvimento Científico e Tecnológico (CNPq), Brazil, for the financial support awarded over the years.

10. References

Bernard, M.C.; Hugot-Le Goff, A.; Joiret, S.; Dinh, N.N. & Toan, N.N. (1999). Polyaniline Layer for Iron Protection in Sulfate Medium. *Journal of the Electrochemical Society*, Vol. 146, No. 3, (March 1999), pp. 995-998, ISSN 0013-4651

Cecchetto, L.; Denoyelle, A.; Delabouglise, D. & Petit, J-P. (2008). A silane pre-treatment for improving corrosion resistance performances of emeraldine base-coated aluminum samples in neutral environment. *Applied Surface Science*, Vol. 254, No. 6, (January 2008), pp. 1736-1743, ISSN 0169-4332.

Conroy, K.G. & Breslin, C.B. (2003). The electrochemical deposition of polyaniline at pure aluminum: electrochemical activity and corrosion protection properties. *Electrochimica Acta*, Vol. 48, No. 6, (February 2003), pp. 721-732, ISSN 0013-4686.

Cook, A.; Gabriel, A. & Laycock, N. (2004). On the Mechanism of Corrosion Protection of Mild Steel with Polyaniline. *Journal of the Electrochemical Society*, Vol. 151, No. 9, (August 2004), pp. B529-B535, ISSN 0013-4651

Coz, F. L.; Arurault, L.; Fontorbes, S.; Vilar, V.; Datas, L. & Winterton, P. (2010) Chemical composition and structural changes of porous templates obtained by anodizing aluminium in phosphoric acid electrolyte. *Surface and Interface Analysis*, Vol. 42, No.1, (February 2010), pp. 227–233, ISSN 1096-9918

Davis, J.R. (1999). *Corrosion of aluminum and aluminum alloys*. ASM International, ISBN 0871706296, Materials Park, OH

DeBerry, D.W. (1985). Modification of the Electrochemical and Corrosion Behavior of Stainless Steels with an Electroactive Coating. *Journal of the Electrochemical Society*, Vol.132, No.5, (May 1985), pp. 1022-1026, ISSN 0013-4651

Dominis, A.J.; Spinks, G.M. & Wallace, G.G. (2003). Comparison of polyaniline primers prepared with different dopants for corrosion protection of steel. *Progress in Organic Coatings*, Vol. 48, No. 1, (November 2003), pp. 43–49, ISSN 0300-9440

Droffellar, H.V. & Atkinson, J.T.N. (1995). *Corrosion and its control: An Introduction to the Subject* (2nd Edition), NACE International, ISBN 1-877914-71-1, Houston

Epstein, A.J.; Smallfield, J.A.O.; Guan, H. & Fahlman, M. (1999). Corrosion protection of aluminum and aluminum alloys by polyanilines: a potentiodynamic and photoelectron spectroscopy study. *Synthetic Metals*, Vol. 102, No. 1-3, (June 1999), pp.1374-1376, ISSN 0379-6779

Fahlman. M.; Jasty, S. & Epstein, A.J. (1997). Corrosion protection of iron/steel by emeraldine base polyaniline: an X-ray photoelectron spectroscopy study. *Synthetic Metals*, Vol. 85, No. 1-3, (March 1997), pp. 1323-1326, ISNN 0379-6779

Fontana, M.G. (1986). *Corrosion Engineering* (3rd Edition), McGraw-Hill Book Company, ISBN 0-07-100360-6, New York

Fujita, J. & Hyland, M.M. (2003). Polyaniline coatings for aluminum: preliminary study of bond and anti-corrosion. *International Journal of Modern Physics B: Condensed Matter Physics, Statistical Physics, Applied Physics*, Vol. 17, No. 8-9, (December 2003), pp. 1164-1169, ISSN 0217-9792

Galkowski, M.; Malik, M.A.; Kulesza, P.J.; Bala, H.; Miecznikowski, K.; Wlodarczyk, R.; Adamczyk, L. & Chojak, M.(2005). Protection of steel against corrosion in aggressive medium by surface modification with multilayer polyaniline - based composite film. *Journal of the Electrochemical Society*, Vol. 150, No. 6, (April 2003), pp. B249-B253, ISSN 0013-4651

Hermas, A.A.; Nakayama, M. & Ogura, K. (2005). Enrichment of chromium-content in passive layers on stainless steel coated with polyaniline. *Electrochimica Acta*, Vol. 50, No. 10, (March 2005), pp. 2001-2007, ISSN 0013-4686

Hintze, P.E. & Calle, L.M. (2006). Electrochemical properties and corrosion protection of organosilane self-assembled monolayers on aluminum 2024-T3. *Electrochimica Acta*, Vol. 51, No.8-9, (January 2006), pp. 1761-1766, ISSN 0013-4686

Huerta-Vilca, D.; Moraes, S.R. & Motheo, A.J. (2003a). Role of a Chelating Agent in the Formation of Polyaniline Films on Aluminum. *Journal of Applied Polymer Science*, Vol.90, No.3, (n.d.), pp. 819–823, ISSN 0021-8995

Huerta-Vilca, D.; Moraes, S.R. & Motheo, A.J. (2003b). Electrosynthesized polyaniline for the corrosion protection of aluminum alloy 2024-T3. *Journal of the Brazilian Chemical Society*, Vol.14, No.1, (January, February 2003), pp. 52–58, ISSN 0103-5053

Huerta-Vilca, D.; Siefert, B.; Moraes, S.R.; Pantoja, M.F. & Motheo, A.J. (2004a). PAni as Prospective Replacement of Chromium Conversion Coating in the Protection of Steels and Aluminum Alloys. *Molecular Crystals and Liquid Crystals*, Vol.415, No.1, (n.d.), pp. 229-238, ISSN 1542-1406

Huerta-Vilca, D.; Moraes, S.R. & Motheo, A.J. (2004b). Anodic treatment of aluminum in nitric acid containing aniline, previous to deposition of polyaniline and its role on corrosion. *Synthetic Metals*, Vol.140, No. 1, (January 2004), pp. 23-27, ISSN 0379-6779

Huerta-Vilca, D.; Moraes, S.R. & Motheo, A.J. (2005) Aspects of polyaniline electrodeposition on aluminium. *Journal of Solid State Electrochemistry*, Vol. 9, No. 6, (n.d.), pp. 416-420, ISSN 1432-8488

Jones, D. A. (1991). *Principles and Prevention of Corrosion* (1st Edition), Macmillan Publishing Company, ISBN 0-02-361215-0, New York

Kamaraj, K.; Sathiyanarayanan, S. & Venkatachari, G. (2009). Electropolymerised polyaniline films on AA7075 alloy and its corrosion protection performance. *Progress in Organic Coatings*, Vol. 64, No. 1, (January 2009), pp. 67-73, ISSN 0300-9440

Kinlen, P.J.; Silverman, D.C. & Jeffreys, C.R. (1997). Corrosion Protection using Polyaniline Coatings Formulations. *Synthetic Metals*, Vol. 85, No.1-3, (March 1997), ISSN 0379-6779

Le, D.P.; Yoo, Y.H.; Kim, J.G.; Cho, S.M. & Son, Y.K. (2009). Corrosion characteristics of polyaniline-coated 316L stainless steel in sulphuric acid containing fluoride. *Corrosion Science*, Vol. 51, No. 2, (February 2009), pp. 330-338, ISSN 0010-938X

Martins, N.C.T.; Moura e Silva, T.; Montemor, M.F.; Fernandes, J.C.S. & Ferreira, M.G.S. (2010). Polyaniline coatings on aluminium alloy 6061-T6: Electrosynthesis and characterization. *Electrochimica Acta*, Vol. 55, No. 10, (April 2010), pp. 3580-3588, ISSN 0013-4686

Moraes, S.R.; Huerta-Vilca, D. & Motheo, A.J. (2002). Polyaniline Synthesized in Phosphate Buffered Media Applied to Corrosion Protection. *Molecular Crystals and Liquid Crystals Science and Technology, Section A: Molecular Crystals and Liquid Crystals*, Vol. 374, No.1, (n.d.), pp. 391-396, ISSN 1058-725X

Moraes, S.R.; Huerta-Vilca, H. & Motheo, A.J. (2003a). Corrosion protection of stainless steel by polyaniline electrosynthesized from phosphate buffer solutions. *Progress in Organic Coating*, Vol.48, No.1, (November 2003), pp. 28-33, ISSN 0300-9440

Moraes, S.R.; Huerta-Vilca, D. & Motheo, A. J. (2003b). Corrosion protection of stainless steel by polyaniline electrosynthesized from phosphate buffer solutions. *Progress in Organic Coatings*, Vol. 48, No. 1, (November 2003), pp. 28-33. ISSN 0300-9440

Moraes, S.R. & Motheo, A.J. (2006). PAni-CMC: Preparation, Characterization and Application to Corrosion Protection. *Molecular Crystals and Liquid Crystals*, Vol.448, No.1, (n.d.), pp. 261/[863]-267/[863], ISSN 1542-1406

Mohseni, M.; Mirabedini, M.; Hashemi, M. & Thompson, G.E. (2006). Adhesion performance of an epoxy clear coat on aluminum alloy in the presence of vinyl and amino-silane primers. *Progress in Organic Coatings*, Vol. 57, No. 4, (December 2006), pp. 307–313, ISSN 0300-9440

Mert, B. D.; Yazici, B.; Tueken, T.; Kardas, G. & Erbil, M. (2011) Anodizing and corrosion behavior of aluminum. *Protection of Metals and Physical Chemistry of Surfaces*, Vol. 47, No. 1, (n.d.), pp. 102-107, ISSN 2070-2051

Mrad, M.; Dhouibi, L. & Triki, E. (2009). Dependence of the corrosion performance of polyaniline films applied on stainless steel on the nature of electropolymerisation solution. *Synthetic Metals*, Vol. 159, No. 17-18, (September 2009), pp. 1903-1909, ISSN 0379-6779

Oliveira, M.A.S.; Moraes, J.J. & Faez, R. (2009). Impedance studies of poly(methylmethacrylate-co-acrylic acid) doped polyaniline films on aluminum alloy. *Progress in Organic Coatings*, Vol. 65, No. 3, (July 2009), pp. 348–356, ISSN 0300-9440

Pathak, S.S. & Khanna, A.S. (2009). Investigation of anti-corrosion behavior of waterborne organosilane-polyester coatings for AA6011 aluminum alloy. Progress in Organic Coatings, Vol. 65, No. 2, (June 2009), pp. 288-294, ISSN 0300-9440

Samui, A.B.; Patankar, A.S.; Rangarajan, J. & Deb, P.C. (2003). Study of polyaniline containing paint for corrosion prevention. Progress in Organic Coatings, Vol. 47, No. 1, (July 2003), pp. 1-7, ISSN 0300-9440

Santos, J.R.; Mattoso, L.H.C. & Motheo, A.J. (1998). Investigation of corrosion protection of steel by polyaniline films. *Electrochimica Acta*, Vol. 43, No. 3-4, (n.d.), pp. 309-313, ISSN 0013-4686

Sathiyanarayanan, S.; Azim, S. S. & Venkatachari, G. (2009). Corrosion protection of galvanized iron by polyaniline containing wash primer coating. Progress in Organic Coatings, Vol. 65, No. 1, (April 2009), pp. 152–157, ISSN 0300-9440

Seré, P.R.; Armas, A.R.; Elsner, C.I. & Di Sarli, A.R. (1996). The surface condition effect on adhesion and corrosion resistance of carbon steel/chlorinated rubber/artificial sea water systems. *Corrosion Science*, Vol. 38, No.6, (June 1996), pp. 853–866, ISSN 0010-938X

Silva, D.P.B.; Neves, R.S. & Motheo, A.J. (2010). Electrochemical Behavior of the AA2024 Aluminum Alloy Modified with Self-Assembled Monolayers/Polyaniline Double Films. *Molecular Crystals and Liquid Crystals*, Vol.521, No.1, (n.d.), pp. 179-186, ISSN 1542-1406

Spinks, G. M.; Dominis, A.J.; Wallace, G. G. & Tallman, D. E. (2002) Electroactive conducting polymers for corrosion control. Part 2. Ferrous metals. *Journal of Solid State Electrochemistry*, Vol. 6, No. 2, (n.d.), pp. 85-100, ISSN 1432-8488

Talboat, D. & Talboat, D. (1998). Corrosion Science and Technology (1st Edition), CRC Press, ISBN 0-8493-8224-6, Boston

Tallman, D.E.; Spinks, G.; Dominis, A. & Wallace, G.G. (2002). Electroactive conducting polymers for corrosion control. Part 1. General introduction and a review of non-ferrous metals. *Journal of Solid State Electrochemistry*, Vol. 6, No. 2, (n.d.), pp. 73-84, ISSN 1432-8488

Talo, A.; Passiniemi, P.; Forsen, O. & Ylaesaari, S. (1997). Polyaniline/epoxy coatings with good anticorrosion properties. *Synthetic Metals*, Vol. 85, No. 1-3, (March 1997), pp. 1333-1334, ISSN 0379-6779

Trethewey, K.R. & Chamberlain, J. (1995). *Corrosion for Science and Engineering* (2nd Edition), Longman Scientific & Technical, ISBN 0-582-238692, England

Trivedi, D.C. (1997). Polyanilines, In: *Handbook of Organic Conductive Molecules and Polymers: Vol.2. Conductive Polymers Synthesis and Electrical Properties*, H. S. Nalwa, pp. 505-572, John Wiley & Sons Ltd, ISBN 0471962759, New York

Zubillaga, O.; Cano, F.J.; Azkarate, I.; Imbuluzqueta, G. & Insausti, M. (2009). Polyaniline and nanoparticle containing anodic films for corrosion protection of 2024T3 aluminum alloy. *Transactions of the Institute of Metal Finishing*, Vol. 87, No. 6, (November 2009), pp. 315-319, ISSN 0020-2967

Part 3

Sensors

3

Application of Conducting Polymers in Electroanalysis

Mohammed ElKaoutit
University of Cádiz
Spain

1. Introduction

Electroanalytical methods or electroanalysis have several advantages over other analytical methods. They offer competitive characteristics such as low cost, fast response, easy automation, miniaturization, portability, the ability to monitor in-situ and in real time, and no need for sample manipulation. They are based on the control of electrical signal and it correlation to analyte concentration. Thus several possibilities of electrochemical transducer have been envisaged but the most commonly used have been divided into the following five categories.

Conductimetric: when a potential is applied across two plates (usually a sine wave voltage), the current that passes through the solution is measured and the conductivity is determined from the current and potential according to the generalized Ohm's law.

Potentiometric: when the potential is measured between the sensing and the reference electrodes under no current flow.

Amperometric: when a constant potential is maintained between a sensing and reference electrodes and current generated by electro-activity of some species at the surface of sensing electrode is measured versus the time.

Voltammetric: which is similar to amperometric, but in this case the potential is varied with time and the current is monitored versus this variation of the potential.

Impedometric: which is based on the simulation of electrochemical cell to electrical circuit, consists of the application of small magnitude perturbation of this circuit with alternative potential, and measuring its impedance. So electrochemical parameters, such as double layer capacitance, resistance solution and impedance of faradaic process, can be determined to characterize the electrode surface and to monitor any change in its state.

In the majority of these approaches the electrode can act as source or captor of electrons transferred from an electroactive molecule located in the interfacial region between the electrode and the solution.

However, this process should not be regarded as perfect. The inertia of the electrode depends on the medium and the applied potential. Many compounds provide similar electrochemical signal (for example; many neurotransmitters exhibit similar values of peaks potential of oxidative process in voltammetric analysis). Some analytes have irreversible and non reproducible oxidoreduction activities at the surface of conventional electrodes, etc... Thus, the electrode must be modified by some recognizing inorganic,

organic, or biological components to detect the analyte with required sensitivity, selectivity and reproducibility.

Development of such so-called heterogeneous electrodes is therefore carried out in order to produce micro-extraction and pre-concentration of analyte, increase the active surface area and working potential window, achieve a good electrical conductivity and inertia, and / or to immobilize biological and chemical recognizing elements. Development of these modified electrodes should be reproducible and controllable in order to be easily transferred to industrial mass production. Also the new surface should be homogeneous, ordered and free of defects to enhance the electron charge transfer and the accessibility of the analyte to recognizing elements or specific sites in the modifier.

Conducting polymers (CPs), which combine the electronic characteristic of metals and inorganic semiconductors, possess the attractive advantage of having easy synthesis control over the properties of the polymeric exposed surface such as structure, morphology and thickness. The use of these polymers represents the most important development in the preparation of modified electrodes to make sensors and biosensors. Their synthesis can be achieved by two conventional methods.

Firstly, chemical synthesis: by adding oxidative agents to a dissolved corresponding monomer in adequate solvent. Thus the electrode must be modified by each one of the popular techniques for depositing controlled thickness film, such as spin-coating, Langmuir-Blodgett-technique or layer-by-layer.

Or secondly by electrochemical method, which has the advantage of combining the synthesis and modification steps in one procedure. So the monomer is electrochemically oxidized at a controlled potential giving rise to free radicals. They are absorbed through the surface of the electrode and subsequently undergo a wide variety of reactions leading to the polymers network. However, the advantage of combining synthesis and modification in a unique procedure should not be considered as exclusively for conducting polymers, because other non-conducting polymers have the same plus point. The difference is that conducting polymers can achieve multilayer and controlled thickness thanks to their intrinsic conductivity.

In addition, these materials were considered as multifunctional in electro-analytical methods development. So they were used as receptors, transducers, immobilization matrix, and/or as anti-fouling and protective materials.

Modification of conventional electrodes with these materials was shown to be an outstanding approach to producing electro-analytical devices able to respond to practically every analytical demand. This approach, and those which were then derived from it, was used to control analytes of interest in environment, agroalimentation, clinical, security terms etc…

Several reviews were published to describe and characterize the advantageous use of conducting polymers especially in the biosensors area (Bartlett & Cooper 1993; Santhanan, 1998; McQuade et al., 2000; Gerard et al., 2002; Cosnier, 2003; 2007; Rahman et al., 2008; Lange et al., 2008; Peng et al., 2009; Nambiar & Yeow, 2011)

In this chapter the advances that have been made in this approach were reviewed, putting special emphasis on the practical aspect and trying to classify the role of CPs in each described device. The topic will be divided into separate sections, each taking into account the analytical characters of this contribution. Thus, we try to provide examples and general reference in the topic for both researchers and postgraduate students.

2. Electrochemical sensor

2.1 Heavy metals control

Heavy metals are common pollutants in the aquatic environment and their hazardous nature is indisputable in our times. Consequently their admissible concentration ranges from zero to a few ppb depending on their toxicity, forms, and the matrix in which they will be determined. Several methods have been developed to control these components with required sensitivity and selectivity. Electrochemical methods have attracted considerable interest in this subject. The most commonly used techniques were based on pre-concentration stripping approaches (either anodic or cathodic), solid phase micro-extraction, and ionic selective electrodes.

Thus, common conducting polymers were used without any modifiers in the pre-concentration / electroanalysis of metal ions, and instead by simple absorption in the heteroatom of CPs, or by its reduction on the surface of the modified electrode and its stripping oxidation control.

Polypyrrole (PPy) micro and macro-electrodes, incorporating chloride ions as the counterion, were grown galvanostatically, and silver ions were pre-concentrated onto the electrode surface for ten minutes from a 1 mol L^{-1} $NaNO_3$ solution. Then the trapped silver was subsequently determined voltammetrically in 1 mol L^{-1} NaCl applying cathodic polarization (Jone & Walace, 1990). A similar approach was optimized by Song and Shiu (Song & Shiu, 2001) achieving detection limit of 0.2 ppm, and using differential pulse cathodic voltammetry, without significant interferences of divalent ions of other heavy metals, especially at low concentration.

However, synthesis of PPy film in the presence of para-sulphonate and its subsequent treatment with 0.5 mol L^{-1} NaOH and 0.5 mol L^{-1} of HNO_3 (base treatment to deprotonation and removal of sulphonate ions and that acidic to return the film to its oxidized state) resulted in the stripping anodic sensor being able to detect 5 ppb of silver ions after its reduction by mercury (Pickup et al., 1998). Modified gold Poly(3-methylthiophene) (P3MeT) was used to detect Hg^{++} using differential pulse voltammetry. The approach was consisted in two steps: first micro-extraction by simple incubating of electro-polymerized P3MeT at open circuit potential for 30 min and then measuring differential pulse voltammetric anodic signal at around pH 2. Low detection limit of 0.1 10^{-9} mol L^{-1} was achieved and the peak of mercury oxidation was not affected by the presence of ions such as Cd^{2+}, Ag^{+}, Fe^{3+}, Cu^{2+}, $Cr_2O_7^{2-}$ and Cr^{3+}(Zejli et al., 2004).

Approaches based on the use of CPs and some molecules (Fig. 1) with notable affinity to heavy metals acting as doping ions were also explored, especially to achieve ionic selective electrodes (ISE).

Electrodes modified by Poly(3,4-ethylenedioxythiophene) (PEDOT) and PPy doped with p-sulfonic calix[4]arene (C4S), p-sulfoniccalix[6]arene (C6S) and p-sulfonic calix[8]arene (C8S) (Fig. 1) were proposed as ISE to silver ions (Mosavi al., 2005). The same strategy was published using PEDOT and several p-methylsulfonated calix[4]resorcarenes (Rn[4]S) with alkyl substitutes of different chain lengths (R1=CH3; R2=C2H5; R3=C6H13) (Fig. 1) was used to detect the same ions (Vázquez et al., 2005a). In these two studies the chemical structure of the backbone of the conjugated polymer seems to play a more important role in the selectivity of potentiometric Ag^{+} sensors based on conducting polymer, than the size of the sulfonated calixarene dopants.

The glassy carbon electrode was polarized in an aqueous solution containing 0.5 mol L^{-1} pyrrole monomer and 0.125 mol L^{-1} Eriochrome Blue-Black B (Fig. 1) as the only added anion, at fixed potential of 0.75 V vs. Ag/AgCl until reaching an optimum electro-polymerization charge of 0.8 mC. This modified electrode was proposed as differential pulse anodic stripping voltammetry and simple potentiometric Ag^+ sensor. For the first method Ag^+ was detected in 0.2 mol L^{-1} KNO_3; pH 2 as electrolyte with low detection limit around 6 10^{-9} mol L^{-1} and the presence of 1000-fold excess of Cd^{2+}, Cu^{2+}, Cr^{3+}, Co^{2+}, Mn^{2+}, Fe^{2+}, Fe^{3+}, Ni^{2+} and Pb^{2+} have been tolerated (Zanganeh & Amini, 2007). Modified glassy carbon electrode with PEDOT doped with hexabromocarborane ($CB_{11}H_6Br_6^-$) (Fig. 1) was also developed by applying a constant charge of 0.014 mA (0.2 mA cm^{-2}) for 714 s to the electrode in the presence of 0.01 mol L^{-1} EDOT and 0.01 mol L^{-1} $AgCB_{11}H_6Br_6$ in acetonitrile, and a resulted modified electrode was used as ISE sensor for Ag^+ (Mosavi et al., 2006). The electrochemical polymerization of PPy was carried out in acetonitrile containing 0.1 mol L^{-1} Tetraphenylborate (Fig. 1) and pyrrole at constant potential of 0.9 V (vs. Ag/AgCl). The film was highly stable and the modified platinum electrode was used as ISE for Zn^{2+} ions (Pandy et al., 2002).

Methods based on electrochemical synthesis and these in turn were based on modifying the electrode with liquid membrane [the membrane consisted of mixture of dissolved poly(3-octylthiophene) (POT) and several other modifiers: potassium tetrakis[3,5-bis(trifluoromethyl)phenyl]borate (KTpFPB), [2.2.2]p,p,p-cyclophane (Fig. 1), and silver hexabromocaborane], were described and compared by Vázquez et al. with the aim of making ISEs for silver. Among these sensors the Nernstian response of Ag^+ was observed for those prepared by electrochemical polymerization of POT only; and those based on thick films of POT doped with the immobile and lipophilic anion $CB_{11}H_6Br_6^-$ (Vázques et al., 2005b).

Bi-polymer of undoped polycarbazole and polyindole was also used as modifier to make ISE to Cu^{2+} (Prakash et al., 2002). The approach was based on consecutive constant potential polymerization of each monomers (polycarbazole at 1.3 V and polyindole at 1V vs. Ag/AgCl) and in removing perchlorate dopant by polarizing at -0.2 V. The undoped ISE achieve detection limit of 10 10^{-6} mol L^{-1} with negligible response to other heavy metals.

The conductometric mercury [II] device was achieved, by the absorption of cryptand-222 (Fig. 1) as a receptor on electro-polymerized polyaniline (PANI), with sensitivity segments in the range of 10^{-12}-10^{-8} mol L^{-1}. The best response of the sensor was observed at pH 2, whereas the authors propose a composite of PANI with surfactants poly(styrene sulphonate) [PSS] and sodium dodecyl sulfate (SDS) to work in neutral media and adapt the method to in-situ analysis (Muthukumar el al., 2007).

Modification of CPs with complexing agent EDTA was also used as a strategy to create height-sensitive sensors for heavy metals. EDTA was covalently attached to electro-synthesized poly(diamino-terthiophene) and extraction of Cu^{2+}, Pb^{2+} and Hg^{2+} was achieved. The detection was carried out in another electrolyte by the reduction of ions at -0.9 V (vs. Ag/AgCl) and re-oxidation by potential scanning from -0.9 to +0.7 V (Rahman et al., 2003). Chemical synthesis of *N*,*N*-ethylenebis[*N*-[(3-(pyrrole-1-yl)propyl) carbamoyl) methyl]-glycine] as pyrrole-EDTA like and its electro-polymerization for the simultaneous determination of Cu^{2+}, Pb^{2+} and Cd^{2+} by the combination of extraction and stripping was reported by Heitzmann et al. Surprisingly, a polymer synthesized in the presence of Cd^{2+} shows more selectivity to this ion (Heitzmann et al., 2007)

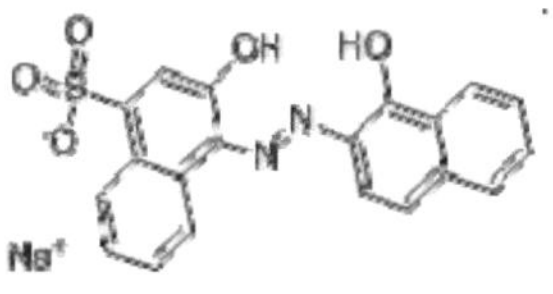

Eriochrome Blue-Black B

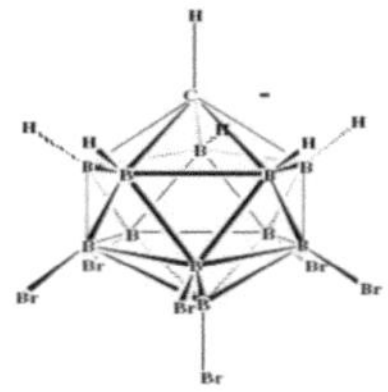

$CB_{11}H_6Br_6^-$

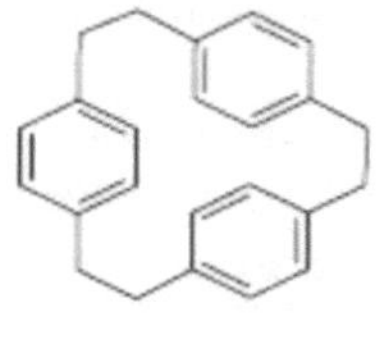

[2.2.2]p,p,p-cyclophane

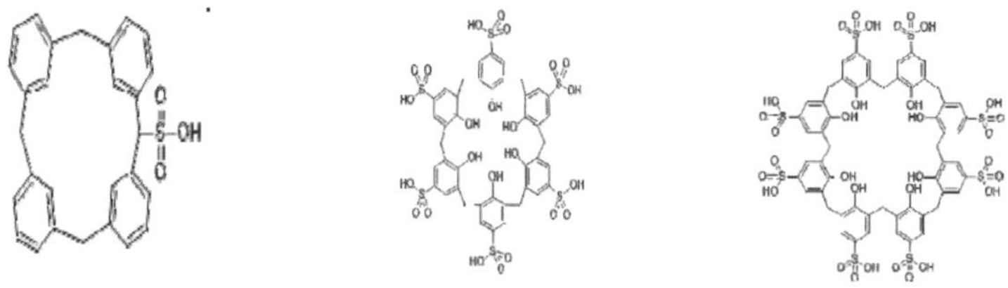

p-sulfonic calix[4]arene (C4S), p-sulfonic calix[6]arene (C6S) and p-sulfonic calix[8]arene (C8S)

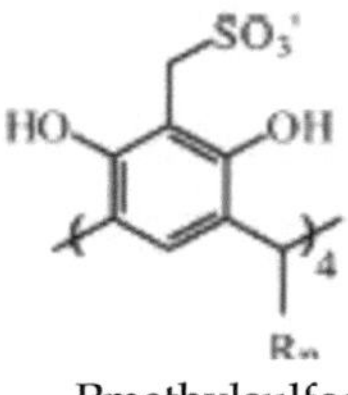

Pmethylsulfonated calix[4]resorcarenes (Rn[4]S) with alkyl substituents of different chain length (R1=CH3; R2=C2H5; R3=C6H13)

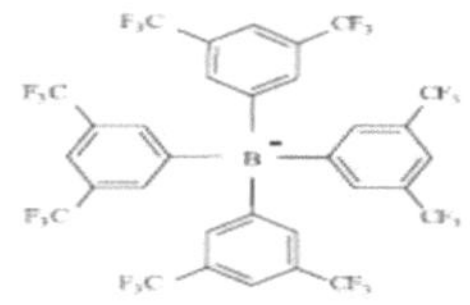

Tetrakis [3,5-bis(trifluoromethyl) phenyl]borate (TpFPB)

Tetraphenylborate

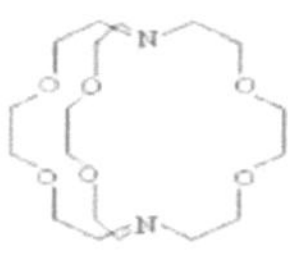

cryptand-222

Fig. 1. Structure of some modifier-dopont used for heavy metals ISE based CPs developments (see text for details)

2.2 Neurotransmitters electroanalysis

Electrocatalytic activity of neurotransmitters and derived forms of pyrogallol and catechol at unmodified conventional electrodes has several disadvantages such as low sensitivity, irreversibility, strong absorption, and hence fouling of the electrode by quinones oxidatives product. Catalysis of their electro-oxidation by the use of conventional CPs, were described in several works. It has been found that oxidation responses of modified P3MeT/Pt electrode to catechol, dopamine, and epinephrine were 4-10 time larger as compared to those of unmodified platinum electrode (Atta et al., 1991). This increase was attributed to the fact that electron transfer occurs at the polymer-solution interface (Mark et al., 1995) not at the inert electrode surface after diffusion through the polymer matrix or through pores (Wang et al., 1989).

Between P3MT, Poly-N-Methylpyrrole, PANI, and Polyfuran, modified polythiophene electrode is the most effective for oxidation of the neurotransmitters (Mark et al., 1995; Atta et al., 1996). Therefore, possibly thanks to sulfur heteroatoms that provides a suitable environment for the electron transfer step for this class of molecules (Kelly et al., 2005).

However in spite of the advantage that CPs have in resolving the previously cited drawback of an unmodified electrode, the resolution of the voltammetric signal of the mixture of catecholamine was not resolved. This was enhanced by the use of micro-electrodes (Galal., 1998; Lupu et al., 2003). But using these electrodes it has been concluded that the peak separation decreases with the increase of monomer concentration, deposition time, and voltage. This is opposite to the achievement of admissible film thickness and electro-modification at -20 °C was proposed as the most acceptable combination of film thickness and electrode properties (Zhang et al., 1997).

Nevertheless, this direct electro-catalysis and current signal magnitudes of CPs modified electrode might be affected by several factors such as electro-synthesis condition, film pre-treatment and the history of the film before its use. The last factor can be considered as inopportune because exposure to some analytes cause a change in film morphology; and the response of this film in proceeding analyses might be totally different to freshly prepared ones.

Electrochemical studding of electro-polymerized P3MeT films by Pd and Pt nanoparticule (NPs) was also explored to achieve stable and selective catalytically CPs based sensors (Atta & El-Kadi, 2009; 2010). It was found that the Pd nanoparticules modified by CPs electrodes (PbNPs, with average diameter of 60 nm , were electro-deposited by cyclic voltammetry method, scanning the potential between −0.25 and +0.65 V (vs Ag/AgCl) at a scan rate of 50 mV^{-1}; 2.5 10^{-3} mol L^{-1} $PbCl_2$ in 0.1 mol L^{-1} $HClO_4$) has the best analytical performance, contrary to that of PtNPs.

Introduction of neutral γ-cyclodextrines into P3MeT doped hexafluorophosphate (Bouchta et al., 2005) and β-cyclodextrines into PPy in the presence of perchlorate (Izaoumen et al., 2005) was described. Cyclodextrines (CDs) are cyclic oligosaccharides consisting of six, seven or eight glucose units called α-, β- and γ-cyclodextrins respectively. These CDs possess a hydrophobic inner cavity and a hydrophilic outer surface. Their well-known ability to form supramolecular complexes with suitable organic and inorganic, neutral, and ionic substances has resulted in the design of selective electrodes. Films growth by cyclic voltammetry from a mixture of monomers and cyclodextrines (100:1 and 20:1 for P3MeT/γ-CD and PPy/β-CD respectively) lead to sensitive surfaces to dopamine, L-dopa and other nuerotransmiters without significant interference of ascorbic acid. A sulfonated β-cyclodextrin was also used as a unique doping ion for PPy film growing at 0.80 V (vs. SCE) in a 0.20 mol L^{-1} pyrrole and 0.01 mol L^{-1} CDs, to achieve a highly selective electrochemical sensor for dopamine, without interference of ascorbate anion, and a detection limits around 3 10^{-6} mol L^{-1} (Harly et al., 2010).

2.3 Moleculary imprinted conducting polymers

Molecular imprinting technology (MIPs) aims to create artificial recognition sites in synthetic polymers. Classically, the process consists of the co-polymerization of monomer and cross linkers in the presence of a template. Thus, the target molecule's "template" is cross linked in the polymer network and the removal or extraction of this template, by adequate solvent, forms binding sites that are complementary in both size and shape.

Solution polymerization, or synthesis, to give small particles with controlled size and physical properties was largely used as the initial process in producing MIPs. This must then be followed by the immobilization of the recognition spheres in close proximity to the electrode; for example through its incorporation into the carbon paste or into supporting gels, as shown in Fig. 2 (Haupt & Mosbach, 2000). The process can also be performed in situ; inside a surface of the sensor, "or transducer", involving imprinted films or membranes. These can be prepared with the deposition of linear polymers in the presence of a template, or by direct polymerization of monomers, cross linkers if it is required, and templates on sensor surfaces, see Fig. 3.

Naturally, the use of conducting polymers to achieve electrochemical sensors based on molecularly imprinted polymers has been utilized in the two ways sited. However thanks to their outstanding intrinsic characteristics theses polymers have also been used as transducers and immobilization matrices.

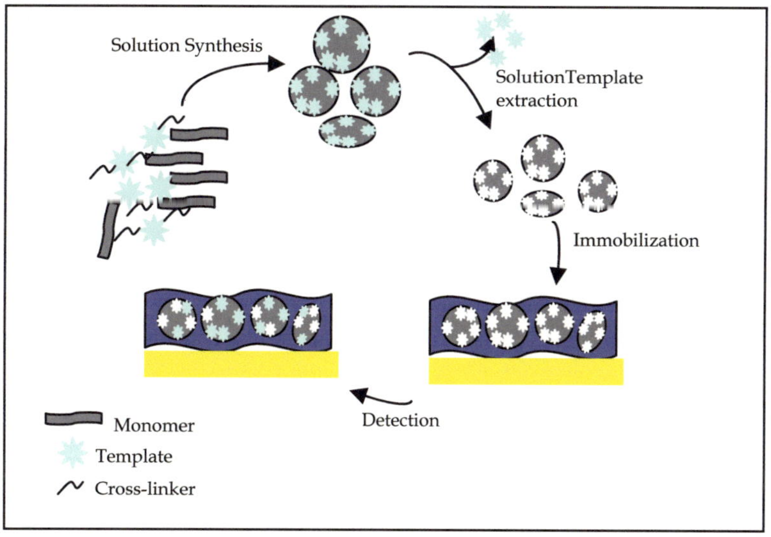

Fig. 2. Solution synthesis of MIPs beads and its immobilization on electrode surface

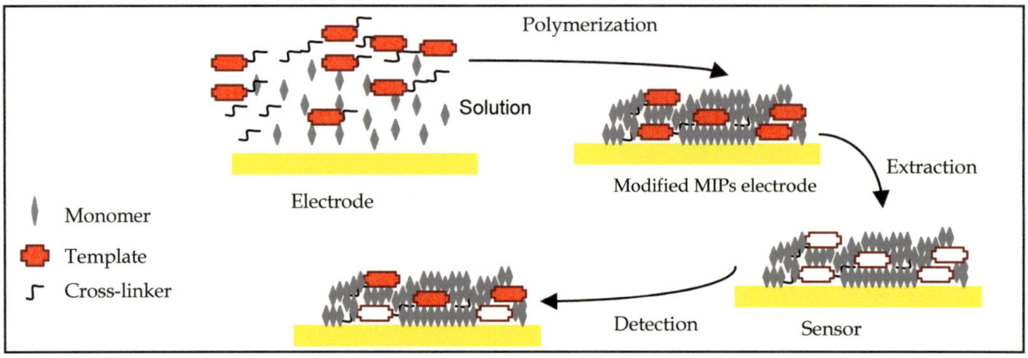

Fig. 3. In-situ synthesis of MIPs

Imprinting polymerization of glucose was performed by the electro-polymerization of poly-o-phenyldiamine (P-*o*-PD) by cyclic voltammetry (20 cycles) in the potential range of 0–0.8 V (vs. Ag/AgCl) at a scan rate 50 mV s^{-1} from a mixture solution of *o*-PD (5 10^{-3} mol L^{-1}) and glucose (0.02 mol L^{-1}) in 0.01 mol L^{-1} acetate buffer (pH 5.18) (Cheng et al., 2001). In this study the authors tried to use this modified electrode as a capacitive sensor, therefore the electrode surface must be well insulated to avoid the penetration of aqueous conducting solution to the layer which in turn increases its capacitance due to the great polarity of water. An insulation process of the electrode consisting in its dipping in 1-dodecanethiol ethanol solution overnight was proposed. The hydrophobic interface was then achieved and the extraction of template process was carried out by simple washing of the electrode. The concentration of glucose was correlated to capacitance (C) values which were calculated directly from the imaginary Z_{im} impedance after recording the current response to an a.c signal with a peak-to-peak amplitude of 10 mV at frequency of 25 Hz. The formula used was:

$$C = 1/2\pi f Z_{im} \tag{1}$$

Voltamperometric MIPs sensors for L-glutamate (L-glu) were achieved thanks to over-oxidation of PPy with dopant complementary cavities. The approach consisted of depositing the film of PPy/L-glu galvanostatically (0.5-0.1 mA cm^{-2}) in aqueous solution containing 0.5 mol L^{-1} monomer and 1.0 mol L^{-1} sodium L-glu as dopant and template for 120-90 min. PPy was over-oxidized potentiodynamically over the range of -0.3 to 1.0 V (vs. Ag/AgCl) at a scan rate of 40 mV s^{-1} at pH 6.9 phosphate buffer. Then the template was removed. Recognition was achieved at pH 1.7 (0.2 mol L^{-1} KCl and 0.2 mol L^{-1} HCl) with the appearance of a pair of peaks at formal potential of +0.2 V attributed to the insertion of cationic form of L-glu in cathodic scan and its rejection in the anodic polarization (Deore et al., 1999).

A similar detection approach was recently used to detect small molecule of atrazine pesticide by Pardieu et al. Film of poly(3,4-ethylenedioxythiophene-co-thiophene-acetic acid) printed with atrazine, due to the hydrogen bonding between the acetic acid substituent and nitrogen heteroatom of atrazine, was synthesized electrochemically. Thus after the association of 3-acetic acid thiophene functional monomer 0.03 mol L^{-1} and atrazine 1.5 mol L^{-1} in CH_2Cl_2 and tetrabutylammonium trifluoromethane sulfonate (TBATFMS as electrolyte and doping ions) for 10 min, 3,4-ethylenedioxythiophene 7.5 mol L^{-1} was added and a constant potential of 1.45 V vs. Pt during 10 min was applied. To remove a template a mixture of protic solvents, methanol/acetic acid solution (0.7:0.3 v/v), was used for 10 min. The sensing process resulted in a cyclic scan of the electrode from -0.5 to 0.4 V, with a scan rate of 25 mV s^{-1}, in CH_2Cl_2 solution containing TBATFMS (0.1 mol L^{-1}). No peak was observed but the author correlated atrazine concentration to relative charge calculated according to:

$$r(Q) = \frac{Q(0M) - Q}{Q(0M)} \tag{2}$$

Where *Q(0M)* is the charge in absence of atrazine and *Q* is in it presence (Pardieu et al., 2009). The mechanism of charge decreasing has not been elucidated by the authors. This could possibly be attributed to a doping/undoping phenomenon, similar to that described by Deore et al. (Diore et al., 1999), but in the inverse sense.

Also, saccharide-imprinted poly(aniline boronic acid) was described involving the selective complexation of boric and boronic acid compounds with saccharides. In this way, the electro-polymerization of derived aniline was achieved at pH (5-7) in the presence of fluoride and the saccharide template was removed by soaking the film in pH 7.4 phosphate buffer solution overnight. A high resolution of potentiometric response of mixture of sacharide such as glucose and fructose was obtained (Deore & Freund, 2003).

On the other hand, micro-spheres with a diameter of 0.5 10^{-6} mol L^{-1} of morphine imprinted polymers were prepared through thermal radical polymerization and the result was immobilized by *PEDOT* conducting polymers into ITO electrode, to be used as amperometric sensors (Ho et al., 2005). Methacrylic acid (MAA) as monomer, and trimethylolpropane trimethacrylate (TRIM) as cross-linker were used. Briefly, 4 ml of MAA and 4 10^{-3} mol L^{-1} of TRIM were mixed in 40 ml of acetonitile, the template morphine was added to be at 57.5 10^{-6} mol L^{-1} and the reaction was initiated by 2,2′-azobisisobutyronitrile. The solution was desoxygenated by sonication for 10 min and then with bubbling nitrogen for 10 min. The breakers were sealed and placed in a water bath at 60 °C for thermal polymerization. After 9 hours the MIPs particles were collected by filtration and the template was removed with methanol. Modification of the ITO electrode was achieved thanks to the electro-polymerization of EDOT in 0.1 mol L^{-1} lithium perchlorate, and 10 mg of MIPs spheres.

3. Electrochemical biosensors

3.1 Classification

A general definition of a biosensor could be given as a chemical sensor in which the recognition system utilizes a biochemical mechanism. However, biosensors must be distinguished from integrated devices which contain a recognition reservoir, an enzymatic or immunological reactor integrated in a flow system to generate products or consume reactive and identify this variation in a separate detection system. A recommended definition of a biosensor is: an analytical device containing a biological recognition element in intimate contact, or immobilized on a surface of phisico-chemical transducer (Thévenot et al., 1999). Biosensors can be classified depending on the type of recognition process (catalytic or affinity), signal transducer (optical, electrochemical…), or immobilization process (cross linking, covalent linkage, adsorption, retention…).

3.1.1 Biocatalytic recognition based on electrochemical sensor

In this case, the biosensor is based on a recognition catalyze reaction. This can be due to the integration of elements in their original biological environment, or in an advanced manipulation process. Thus, three possible types of biocatalysts entity were envisaged:

a. Whole cells including micro-organisms, such as bacteria, fungi eukaryotic cells or yeast, and cells organelles such as mitochondria and cells walls.
b. Tissue.
c. Enzyme (mono or multi-enzyme) which can be directly isolated and purified from natural microorganisms or engineered to achieve recognition elements with specific characteristic.

The general reaction scheme for biocatalytic process can be expressed as:

$$S + S' \xrightarrow{Biocatalyst} P + P'$$

Where one or more analytes, usually named substrates, S and S', react in the presence of enzyme(s), whole cells or tissue culture and yield one or more products (P and P').
There are three possibilities that use adjacent transducers for monitoring this biocatalysed reaction. These possibilities have been used as chronological classification of the electrochemical biosensors historical evolution, and can be categorized as:

3.1.1.1 First generation biosensors

First generation biosensors rely on the measurement of decreasing initial signal values of co-substrate e.g. oxygen depleted by oxidase, bacteria or yeast reacting layers; or the increase of products such as hydrogen peroxide, quinone, H^+, CO_2, or NH_3, etc. For example if enzyme glucose oxidase is used, glucose concentration can be monitored as a result of either oxygen consumption (the historical Clark and Lyon biosensor was based on this approach), or of peroxide generation following the reaction and corresponding scheme in Fig.4:

$$Glu\cos e + O_2 \xrightarrow{GOX} Glucoonolactone + H_2O_2$$

3.1.1.2 Second generation biosensors

In the last example, hydrogen peroxide was controlled on the surface of the working electrode (a Pt electrode usually functions as the anode at +700 mV). This includes both an oxidation process at high potential and species such as ascorbate, urate, etc., that may electrochemically interfere. Chemically synthesized mediators, usually molecules foreign to the natural catalytic cycle of enzymes, were introduced to the systems in order to react firstly with a product enzymatic reaction - especially peroxide - and then with the enzyme redox centers. Thus the second generation biosensor is defined.

3.1.1.3 Third generation biosensor

In general this class of biosensor is limited to redox proteins and enzymes with one or more electroactive cofactors tightly bound to protein molecules, or integrated in other subunits. Accordingly, this type of biosensor consists of control of direct electronic transfer between these cofactors and the electrode surface. So, following the oxidation or reduction of the substrates, the redox state of active centre of the enzymes is changed, and the electrons are transferred to the electrode or to another acceptor/donor cofactor integrated in subunits of proteins which must ultimately be transferred to the electrode too. Flavin-adenine-dinucleotide (FAD^+), nicotinamide-adenine-dinucleotide (NAD^+), Nicotinamide adenine dinucleotide phosphate ($NADP^+$), pyrrolo-quinoline-quinine (PQQ), heme, and other metallic derivative centres such as cooper, molybdenum, or iron-sulfur, and the combination of same elements, are the most frequently cofactor units studied in electrochemical biosensors.

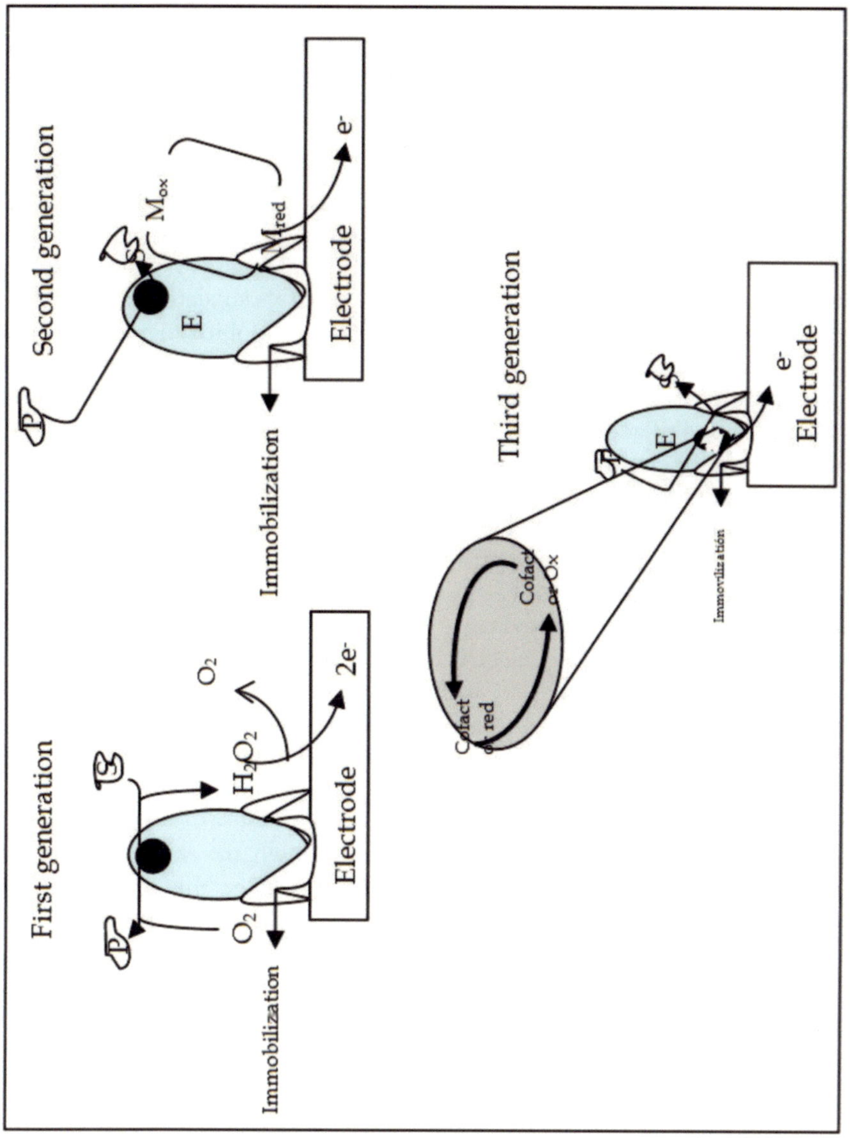

Fig. 4. Three generations of electrochemical biosensors based biological catalyst

3.1.2 Bioaffinity recognition based electrochemical sensor

In this class of biosensor the analyte interacts with a specific site in the biomolecule, or organized molecule assemblies, which have been isolated from the biological environment or synthetically engineered. After this interaction the equilibrium was reached, the analyte is no further consumed by the immobilized recognising agent, but the achieved equilibrium is monitored directly from continuous control of electrochemical surface parameters or indirectly using a complementary biocatalytic reaction. Steady-state or transient signals are then monitored by the integrated detector. The most commonly developed strategies are based on: immunological specific interaction between an isolated or engineered antibody and analyte antigen (immunoelectrochemical sensor), specific DNA fragments interaction (DNA sensor), and interaction between synthesized oligonucleotide and target analyte (aptasensor).

3.1.2.1 Immunoelectrochemical sensor

Antibodies are proteins produced in animals as an immunological response to the presence of a foreign substance called antigen, and have specific affinity for this antigen. The first categories of immunoelecrochemical sensor was based on the existing schemes of enzymatic immunoassays ELISA (Enzyme Linked Immune Sorbent Assay) which usually need a labelled antibody or antigen, and is classified in two generals subcategories, as shown in Fig. 5. In the non-competitive immunoassay, the antibody is immobilized and, after the addition of the sample which contains the antigen, a conjugated or secondary labelled antibody is added. In competitive assays, the competition can either be between the free antigen (from the sample) and immobilized antigen for limited controlled amount of labelled antibody (complexion of all antigens from the sample and correlation of its concentration with the excess of labelled antibodies); or between the antigen and labelled antigen for limited immobilized amount of antibodies, Fig. 5. Labelling of antigen or antibody is usually achieved using oxido-reductase enzyme, such as alkaline phosphatase, horseradish peroxidase, or glucose oxidase. However the use of direct electro-active chemical compound and metallic nanoparticle was also envisaged.

The second category is label-free electrochemical immunosensor using electrochemical impedance spectroscopy (EIS) and/or volammretry. These have been explored widely due to their high sensitivity and they are based on detecting the electron transfer of foreign redox pairs (usually added to the measure solution), or change in capacitive double layer for non- faradaic processes after antibody-antigen recognition.

For limited example of antibody and antigen couples, label-free immunosensor can be also achieved as a result of the control of current given by electro-active residues in immobilized antibody, thereafter the current changes after the binding of the target antigen with the antibody as well as with the binding of secondary antibody. This process based on protein electro-activity is amplified by the use of carbon nanotube (Vestergaard et al., 2007).

3.1.2.2 DNA hybridization and aptamers based sensors

These sensors rely on the immobilization of a relatively short single standard sequence of natural DNA or synthesized oligonucleotide on the electrode surface. The interaction with a specific complementary region of the target DNA gives rise to a DNA hybridization sensor. The interaction of the oligonucleotide with a target molecule such as a protein drug or environmental effluent gives apatmers biosensors.

Electrochemical signal transduction in these types of biosensors is similar to that of immunosensors. This can be based on; 1) labelling the target or the immobilized oligonuclotide, 2) directly intercalating into the duplex, during the formation of a double-stranded DNA on the probe surface, a molecule that is electroactive, 3) or direct detection of guanine which is electroactive too. Control of the electron transfer of redox pairs, or impedance measurement of non faradaic process by EIS or CVs was also envisaged.

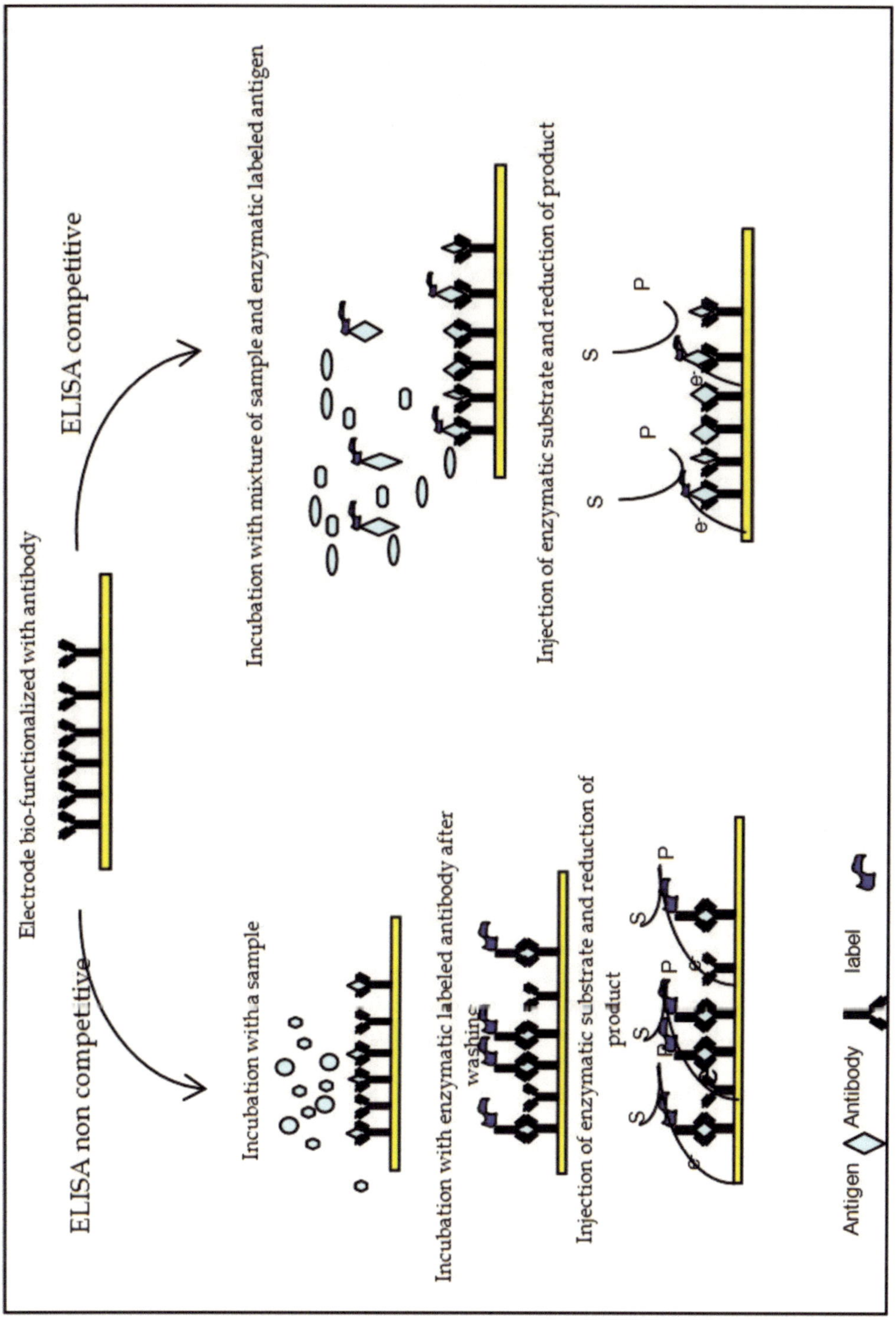

Fig. 5. Configuration of ELISA competitive and non competitive immunosensor

3.2 CPs based electrochemical biosensor

From the definition of a biosensor, development of these devices must undergo three steps: adequate choice of biological recognition element and its modification, its immobilization on an electrode surface, and designs of transduction process. The use of conducting polymers in this area essentially takes place in the two last steps. CPs are therefore used as immobilization matrices especially when their electro-polymerization property is taken into consideration, and as transducers or sensing probes due to their charge donors and acceptors characteristic.

3.2.1 Use of CPs as immobilization matrix

Methods of immobilization of biological recognition element are categorized as physical and chemical. Physical immobilization occurs without any modification of biological element and was based on electrostatic or hydrophobic interaction of these elements and modified electrode, or on simply retention - with membrane - on the top of a metallic electrode. Chemical immobilization was based on covalent binding of the biological element to a modified electrode. The use of CPs as a basis of an immobilization matrix was explored in these two strategies, and also on the legitimised approach to their versatile electropolymerization.

3.2.1.1 Electro-immobilization

Electro-immobilization of biological recognition elements can be achieved in a simple one-step method involving the application of appropriate potential to the working electrode, soaked in a mixture of electro-polymerizable monomer and bimolecules. The mechanism of such immobilization was reviewed by Bartlett and Cooper (Bartlett & Cooper, 1993), and it has been found on the charge complementary between the CPs and the enzyme. Accordingly, since the CPs is deposited as a polycation, electrostatic interaction between the polymers and the biomolecule occurs and the enzyme is then retained in the three dimensional growth film. This mechanism was supported by three pieces of evidence: First, simple adsorption of the bio-molecule onto performed CPs film controlling the net charge of each component (CPs are positively charged in their oxidized form and biomolecules are negatively charged at pH above their isoelectric point). Secondly, the amount of incorporated enzyme and the enzymatic activity of the biofilm decrease, as the concentration of other small anions increases. And thirdly, that it is difficult to incorporate positively charged biomelecules into CPs. However this mechanism has been rebuffed by Schumann who proposes that the entrapment of an enzyme is only due to a statistical enclosure of the enzyme present in the vicinity of the electrode surface (Schumann et al., 1993).

In any case this method has several disadvantages, including the difficulty of co-immobilization of enzyme and mediator, due to their competition during the charge compensation process; the high isoelectric point of some biomolecules which can be incompatible with polymer pH polymerization; and the requirement of high enzyme concentration.

This approach was enhanced by Cosnier and Innocent, using synthesized amphiphilic pyrrole monomers (Cosnier & Innocent, 1992). A monomer and enzyme-saving technique for biosensor preparation has been developed based on initial absorption of the enzyme together with the amphiphilic pyrrole derivative on the electrode surface, followed by solvent evaporation and subsequent electrochemical polymerization in aqueous electrolyte solution. Enzyme pre-absorption onto the electrode surface prior to the initiation of electro-polymerization process can be also an enhancement of electro-immobilization. Accordingly, if the polymer/enzyme film which is subsequently grown is relatively thin, this will increase the overall concentration of the enzyme and may enhance the sensitivity of the resulting biosensor.

3.2.1.2 Covalent binding

This type of binding includes two steps, activation of a solid matrix and binding of the biomolecule or its derivative forms. Modification of electrode surface with deposited conducting polymers introduces the apparent possibility for activation of these electrodes.

Already activated amine or carboxylic monomer can be polymerized or copolymerized with an unsubstituted parent to give a directly activated surface, which can be covalently biomodified using carbodiimide. Various N- and 3-substituted pyrrole monomers have been synthesized in this order. The synthesized pyrrole derivatives were tested by Schalkhammer et al., for their ability to form polymers as well as to retain the GOX enzyme activity and permeability of peroxide product to electrode surface. The authors recommend the use of co-polymerization of substituted and unsubstitued pyrrole as an efficient way to achieve aqueous stability and reactive polymer films. The most promising substituted pyrrole was 2-(1-pyrrole)-acetylglycine which forms only thin water-soluble polymer films, but can be co-polymerized with pyrrole to obtain porous films with an optimal enzyme load. Also COOH- and NO_2-groups (NO_2 group was reduced after electro-polymerization of corresponding monomer, using a solution of 15% TiC13 or 1% SnCl, in l% HCl for 30 min to form a film with pending NH_2 for enzyme linking), which are stable against oxidation under polymerizing conditions have proved to be optimal for obtaining a modified PPy layer which can be used for covalent coupling of enzymes (Schalkhammer et al., 1991).

Unsubstituted polymer film can be derived from a heterogeneous reaction on the electrode surface. The first example was reported by Schuhmann et al., it is based on nitration of electrogrown PPy film and covalent bending of glucose oxidase. Thus, after electrochemical polymerization of PPy in water and oxygen free 0.1 mol L^{-1} solution of pyrrole in acetonetrile containing 0.1 mol L^{-1} Tetrabutylammonium p-toluenesufonate (TBAPTS). The polymer was nitrated by keeping the modified electrode in a solution of 700 mg $Cu(NO_3)_2.3H_2O$ in 20 mL acetic anhydride for 5 min at 20° C in an Ar atmosphere. The nitro groups were then reduced electrochemically in CH_3CN/TBAPTS, cycling the potential between 500 and -250 mV vs. SCE. Carboxylic side chains of the GOX enzyme were activated with water soluble cabodiimide and the result was covalently linked to such in-situ synthesized poly(3-aminopyrrole) (Schuhmann et al., 1990).

A comparable possible covalent approach could be the binding of a monomer to the enzyme. There are no differences found in the oxidation potential of a simple pyrrole and that of a glucose oxidase-bound pyrrole. Consequently, subsequent polymerization of the protein-functionalized monomer with a non-functionalized gives film with controlled enzyme concentrations, and provides greater sensitivity; possibly arising from the greater porosity of the film. As an example pyrrole-modified glucose oxidase with pyrrole units can be obtained via amide bonds or secondary amines. A stable modified enzyme-pyrrole with an average of 30 pyrrole units was obtained thanks to the formation of amide bonds between caboiimide-derivated N-(2-caboxyethyl)pyrrole derivatives and lysyl residues at the surface of protein (Wolwacz et al., 1992). Similarly, N-(3-aminopropyl)pyrrole has been bound to carbodiimide-activated carboxylic residue of the enzyme (Yon-Hin et., al 1993). A similarly obtained glucose-functionalized pyrrole monomer was also polymerized with bithiophene to achieve bio-polymers with higher redox potential compared to that of pyrrole; this assures film with more stability toward degradation effect of peroxide product of GOX (Hiller et al., 1996).

The electrochemical polymerization of monomers functionalized by easy leaving groups such as N-hydroxysuccinimide or N-hydroxyphthalimide has also been described as simple post-polymerization functionalization of CPs. The precursor copolymer of poly[(3-acetic acid pyrrole)/3-N-hydroxyphthalimide pyrrole] was electro-copolymerized at 0.9 V (vs. SCE) onto a platinum electrode in acetonitrile and an amino-substituted oligonucleotide was then grafted onto it by direct chemical substitution of the leaving group N-hydroxyphthalimide, in dimethylformamide containing 10% acetate buffer at pH 6.8 during 3 h incubation time (Korri-Youssoufi et al., 1997). The conversion of (5-bromopentyl)-substituted mono, bi and tri-thiophene, to carboxylic acid was accomplished in two steps; first with KCN, H_2O/EtOH and second in KOH, H_2O/MeOH. The carboxylic acid was further converted to the acid chloride by oxalyc chloride and directly reacted with N-hydroxysuccimide (NHS) and triethylamine to give the NHS-ester. A comparison of the electrochemical polymerization process of these thiophenes substituted with activated NHS-ester resulted in an oxidative potential more positive for monothiophene, compared to bi or tri-thiopene which both show similar potential (Bäurele et al., 1996).

Another means of obtaining functionalized CPs films is through electro-immobilization of conventional carboxylic or amine polymers with a CPs monomer. For example, electro-generation of polyaniline/poly(acrylic acid (PANI/PAA) films were polymerized on Boron Doped Diamond electrode (BDD) in 0.1 mol L^{-1} H_2SO_4/0.5 mol L^{-1} Na_2SO_4 solution containing 0.2 mol L^{-1} aniline monomer and 25 mg mL^{-1} poly(acrylic acid) (MW=2000) by potential cycling one time from -0.2 V to +1.2 V (vs. Ag/AgCl) at various scan rates. The carboxylic group density of this film was quantified using the Toluidine Blue O (TBO) method, which consisted of soaking the modified electrode in an aqueous solution of 0.5 10^{-3} mol L^{-1} TBO, adjusted to pH 10 with NaOH. The formation of ionic complexes between the carboxylic acid groups of the PANI/PAA film and cationic dye was allowed to proceed for 5 h at room temperature. After washing the electrode with the same alkaline water solution to remove uncomplexed TBO, the dye was desorbed by 50wt % acetic solution and the amount of carboxylic group was calculated from the optical density of the desorbed dye at 633 nm assuming that 1 mole of TBO had combined with 1 mole of carboxylic group of the PAA. In this study it was noted that such an increase in scan rate during electropolymerisation of PANI/PAA films had a largely negative effect on the carboxylic density of the film; therefore the optimum condition was found to be at a scan rate of 10 mV s^{-1} to attain ultrathin film with the highest carboxylic functionalities (Gu et al., 2005).

3.2.1.3 Affinity interaction with functionalized CPs films

The sensitivity of the biosensor was greatly dependent on the orientation of immobilized biological recognition elements. Chemical binding of a bio-molecule does not fulfill this criterion and therefore cannot ensure the biological activity preservation. However, affinity interaction between functionalized CPs films and bio-molecules, which induces regioselectives grafting was regarded as a convincing method to achieve high oriented bio-surface with required sensitivity.

Specific and high-affinity interaction between biotin and avidin (association constant K_a= 10^{15} mol L^{-1}) leads to strong associations similar to the formation of covalent bonding. This

high and specific linkage was largely valued for binding biological species to the surfaces. Using conducting polymers, this strategy involves the electropolymerization of conducting monomer-modified biotin (Cosnier et al., 1999) or electro-immobilization of streptavidin (Xiao et al., 2007), the resulting conducting polymer is subsequently made operational by avidin (for the first case) and finally by biotinylated bio-molecules. This strategy comes with some distinct advantages, such as availability of several biotinylated bio-molecules - especially antigen and antibody-, a surface modified with avidin provides a passivated interface that prevents further non-specific adsorption of proteins, and also the high accessibility of the resulting immobilized molecule.
Another strategy is based on the formation of transition metal complexes between histidine tagged proteins, where the surface is made functional by immobilized nitrilotriacetic or iminodiacetic (Haddour et al., 2005).

3.2.2 Use of CPs as transducer

In spite of providing several easy means of actually immobilizing bio-molecules, CPs are a good energy transducer thank to their redox activity and efficient protectors of the electrodes against interfering materials. CPs are able to transduce directly the electrochemical signal generated by some redox enzymes; sense DNA hybridization; and detect directly antibody-antigen interaction. They are also capable of making credible third generation, DNA, and immunological biosensors, respectively. The following examples illustrate some representative reports in these areas.
Glucose oxidase considered as a typical flavin enzyme with FAD/FADH as redox prosthetic group was largely studied to contrast the ability of using some material to be the base of third generation biosensor development. In this trial, the nano-structured or thin film of CPs polyaniline and poly(3,4-ethylenedioxythiophene), was seen to promote the direct electron transfer of immobilized GOX.
In their comparative study Thomson et al. show that contrary to GOX electro-immobilized with poly(3,4-ethylenedioxythiophene), the enzyme absorbed onto a vapor phase polymerized PEDOT shows strong peaks attributable to FAD redox activity with average potential at -0.505 mV vs. Ag/AgCl. Vapor deposition of PEDOT was achieved by simple introduction of gold electrode modified with a thin layer of 40% iron(III) para-toluene sulfonate in butanol and 2.3% pyridine in a sealed EDOT monomer chamber at 70 °C for 30 min (Thompson et al., 2010).
Zhao et al. optimized the synthesis of polyaniline nanofibers and relate its application to direct electron transfer of glucose oxidase. The synthesis of polyaniline nanofibers was carried out at an interface of toluene which contains 0.1 mol L^{-1} aniline and aqueous 1 mol L^{-1} sulfuric acid containing 0.05 mol L^{-1} of ammonium peroxydisulfate. The resulting two-phase system was left undisturbed at room temperature. After 30 s polyaniline appeared at the interface and initiated its migration into the aqueous phase. At this point the phase becomes dark-green and stops changing for 6 h, which was considered as sufficient reaction time. Finally the polyaniline nanofibers were collected by filtering and washing them with water and ethanol and subsequently vacuum-dried. For manufacture of the modified electrode the nano-fibres were suspended in aqueous solution, and then deposited onto the surface of a glassy carbon electrode. After the evaporation of the solvent, the electrode was covalently bio-modified with GOX using EDC-NHS mixture. Contrary to the electrode without GOX, the cyclic voltammograms of these manufactured biosensors showed redox

peaks with formal potential at -418 mV (vs. Ag/AgCl) and a catalytic response to glucose at -0.350 mV (Zhao et al., 2009). This was characteristic of direct electron transfer of FAD/FADH as a redox center in the enzyme. Similar values were obtained by Wang et al. when GOX was electro-immobilized onto the inner wall of highly ordered polyaniline nano-tubes, which were then synthesized using anodic aluminium oxide membrane as a template (Wang et al., 2009).

The use of a template to achieve nano-structured PEDOT was also explored by Arter et al. A nickel nano-trench was lithographically patterned to create devices with several hundred linear arrays on glass. This device was used as the working electrode in the three electrode cells for simultaneous electro-deposition of virus-M 13 and PEDOT after initiation by a pure PEDOT "primer" layer. The resulting device was used as a resistive biosensor for positive antibody (p-Ab). The electrochemical resistivity of the device increases with increasing concentration of p-Ab and a charge-gating was postulated as the mechanism of this increase after p-Ab/M 13 virus affinity binding (Arter et al., 2010).

Conductimetric immuno-biosensors were also reported by Mohammed-Tahir and Alocilja. Their approach can be expressed as 'sandwich' methodology: using a secondary antibody conjugated by polyaniline to achieve highly sensitive *E-coli* and *Salmonella* biosensors. In this manner, the primary antibody was immobilized by a simple glutraldehyde method, the antigen was incubated with this functional surface and a secondary antibody was used to close the electrical circuit between two lateral inert electrodes (Mohammed-Tahir & Alocilja, 2003)

Label-free affinity biosensors using CPs as a transducer were characterized by Sadik et al. The modification of polypyrrole with anti-human serum albumin (anti-HSA) was achieved by simple-step electrochemical immobilization, and the interaction of antigen (HSA) and antibody-PPy electrode was monitored using cyclic voltammetry and impedance spectroscopy at different potentials, corresponding to reduced, doped and overoxidized states of PPy. Impedance measurement showed that large values of double layer capacitance were observed at higher positive potentials with a transition from kinetic control to diffusion control (Sargent et al., 1999). A similar result was observed when anti-bovine serum albumin was electro-immobilized with PPy film and detectable antibody-antigen interaction was measured by the control of the Faradaic behavior of the electrode (Grant et al., 2005).

Change of the doping/undoping redox reaction of CPs to control hybridization of DNA was reported by Mohan et al. Accordingly, chemical immobilization of ssDNS onto electrochemically synthesized poly 5-carboxilic indol, to achieve sensitive biosensor for breast cancer was reported by the plot of change in charge resistance or capacitance with log concentration of target DNA (Mohan et al., 2010).

4. Conclusion and perspective

In this chapter we have summarized the advantageous use of conducting polymers for electrochemical sensing and bio-sensing. These advantages lie predominantly in surface modification controls and electrochemical redox activity which make these materials an useful transducer. Some examples of nano-structured CPs and their use in promoting direct electron transfer, as well as third generation biosensor, were also outlined. The points covered could well be the future focal point of this topic; as well as that of the characterization of signal transduction in label-free bio-affinity biosensors based conducting polymers.

5. Acknowledgment

The author is grateful to Prof. Dr. José Luís Hidalgo-Hidalgo de Cisneros and Prof. Dr. Ignacio Naranjo Rodriguez for their scientific revision of the manuscript. The Junta de Andalucía/FEDER (FQM249 / P08-FQM-04006) and Ministerio de Ciencia e Innovación of Spain/FEDER (CTQ2007-67563/BQU) are also acknowledged for their financial support.

6. References

Arter, J.A; Taggart, D.K; McIntire, T.M; Penner, R.M; Weiss, G.A. (2010). Virus-PEDOT nanowires for biosensing. *Nano Letters*, Vol.10, No.12, (December 2010), pp. 4858-4862, ISSN 1530-6984

Atta, N.F; El-Kady, M.F. (2009). Poly(3-methylthiophene)/palladium sub-micro-modified sensor electrode. Part II: Voltammetric and EIS studies, and analysis of catecholamine neurotransmitters, ascorbic acid and acetaminophen. *Talanta*, Vol.79, No.15, (August 2009), pp. 639-647, ISSN 0039-9140

Atta, N.F; El-Kady, M.F. (2010). Novel poly(3-methylthiophene)/Pd, Pt nanoparticle sensor: Synthesis, characterization and its application to the simultaneous analysis of dopamine and ascorbic acid in biological fluids. *Sensors and Actuators B*, Vol.145, No.1, (March 2010), pp. 299-310, ISSN 0925-4005

Atta. N.F; Galal, A; Ozler, A.E.K; Russell, G.C; Zimmer, H; Mark, Jr.H.B. (1991). Electrochemistry and detection of some organic and biological molecules at conducting poly(3-methylthiophene) electrodes. *Biosensors and Bioelectronics*, Vol.6, No.4, (not found), pp. 333-341, ISSN 0956-5663

Atta, N.F; Marawi, I; Petticrew, K.L; Zimmer, H; Mark, Jr. H.B; Galal, A. (1996). Electrochemistry and detection of some organic and biological molecules at conducting polymer electrodes. Part 3. Evidence of the electrocatalytic effect of the heteroátomo of the poly(hetetroarylene) at the electrode/electrolyte interface. *Journal of Electroanalytical Chemistry*, Vol.408, No.1-2, (May 1996), pp. 47-52, ISSN 0022-0728

Bäuerle, P; Hiller, M; Scheib, S; Sokolowski, M; Embach, E. (1996). Post-polymerization functionalization of conducting polymers: Novel poly(alkylthiophene)s substituted with easily replaceable activated ester groups. *Advanced Materials*, Vol.8, No.3, (March 1996), pp. 214-218, ISSN 0935-9648

Bartlett, P.N; Cooper, J.M. (1993). A review of the immobilization of enzyme in electropolymerized films. *Journal of Electroanalytical Chemistry*, Vol.362, No.1-2, (December 1993), pp. 1-12, ISSN 0022-0728

Bouchta, D; Izaoumen, N; Zejli, H; El Kaoutit, M; Temsamani, K.R. (2005). A novel electrochemical synthesis of poly-3-methylthiophene-γ-cyclodextrin film Application for the analysis of chlorpromazine and some neurotransmitters. *Biosensors and Bioelectronics*, Vol.20, No.11, (May 2005), pp. 2228–2235, ISSN 0956-5663

Cheng, Z; Wang, E; Yang, X. (2001) Capacitive detection of glucose using molecularly imprinted polymers. *Biosensors and Bioelectronics*, Vol.16, No.3, (May 2001), pp. 179–185, ISSN 0956-5663

Cosnier, S; Innocent, C. (1992) A novel biosensor elaboration by electropolymerization of an adsorbed amphiphilic pyrrole-tyrosinase enzyme layer. *Journal of electroanalytical Chemistry*, Vol.328, No.1-2, (July 1992), pp. 361-366, ISSN 0022-0728

Cosnier, S; Stoytcheva, M; Senillou, A ; Perrot, H; Furriel, R.P.M; Leone, F.A. (1999). A biotinylated conducting polypyrrole for the spatially controlled construction of an amperometric biosensor. *Analytical Chemistry*, Vol.71, No.17, (September 1999), pp. 3692-3697, ISSN 0003-2700

Cosnier, S. (2003) Biosensors based on electropolymerized films: New trends. *Analytical and Bioanalytical Chemistry*, Vol. 377, No. 3, (October 2003) pp. 507–520, ISSN 1618-2642

Cosnier, S. (2007). Recent advances in biological sensors base don electrogenerated plymers: A review. *Analytical Letters*, Vol. 40, No. 7, (Not found), pp. 1260-1279, ISSN 0003-2719

Deore, B; Chen, Z; Nagaoka, T. (1999). Overoxidized polypyrrole with dopant complementary cavities as a new molecularly imprinted polymers matrix. *Analytical Science*, Vol.15, No.9, (September 1999), pp. 827-828, ISSN 0910-6340

Deore, B; Freund, M. S. (2003). Saccharide imprinting of poly(aniline boronic acid) in the presence of fluoride. *Analyst*, Vol.128, No.6, (June 2003), pp. 803–806, ISSN 0003-2654

Galal, A. (1998). Electrochemistry and characterization of some organic molecules at "Microsize" conducting polymer electrodes. *Electroanalysis*, Vol.10, No.2, (February 1998), pp. 121-126, ISSN 0022-0728

Gerard, M; Chaubey, A; Malhotra, B.D. (2002). Application of conducting polymers to biosensors. *Biosensensors and Bioelectronics*, Vol.17, No.5, (May 2002), pp. 345–359, ISSN 0956-5663

Gu, H; Su, X.d; Loh, K.P. (2005). Electrochemical impedance sensing of DNA hybridization on conducting polymer film modified diamond. *The Journal of Physical Chemistry B*, Vol.109, No. 28, (July 2005), pp. 13611-13618, ISSN 1089-5647

Grant, S; Davis, F; Law, K.A; Barton, A.C; Collyer, S.D; Higson, S.P.J; Gibson, T.D. (2005). Label-free and reversible immunosensor based upon an ac impedance interrogation protocol. *Analytica Chimica Acta*, Vol.537, No.1-2, (April 2005), pp. 163-168, ISSN 0003-2670

Haddour, N; Cosnier, S; Gondran, C. (2005). Electrogeneration of poly(pyrrole)-NTA chelator film for reversible oriented immobilization of histidine-tagged proteins. *The Journal of the American Chemical Society*, Vol.127, No.16, (April 2005), pp. 5752-5753, ISSN 0002-7863

Harley, C.C; Rooney, A. D; Breslin, C. B. (2010). The selective detection of dopamine at a polypyrrole film doped with sulfonated β-cyclodextrins. *Sensors and Actuators B*, Vol.150, No.2, (October 2010), pp. 498-504, ISSN 0925-4005

Haupt, K; Mosbach, K. (2000). Moleculary imprinted polymers and thier use in biomimetic sensors, *Chemical Reviews*. Vol.100, No.7, (July 2000), pp. 2495-2504, ISSN 0009-2665

Heitzmann, M; Bucher, C; Moutet, J-C; Pereira, E; Rivas, B.L; Royal, G; Saint-Aman, E. (2007). Complexation of poly(pyrrole-EDTA like) film modified electrodes: Application to metal cations electroanalysis. *Electrochimica Acta*, Vol.52, No.9, (February 2007), pp. 3082–3087, ISSN 0013-4686

Hiller, M; Kranz, C; Huber, J; Bäuerle, P; Schuhmann, W. (1996) Amperometric biosensors produced by immobilization of redox enzymes at polythiophene-modified electrode surfaces, *Advanced materials*, Vol. 8, No.3, (March 1996), pp. 219–222, ISSN 0935-9648

Ho, K-C; Yeh, W-M; Tung, T-S; Liao, J-Y. (2005). Amperometric detection of morphine based on poly(3,4-ethylenedioxythiophene) immobilized molecularly imprinted polymer particles prepared by precipitation polymerization. *Analytica Chimica Acta*, Vol.542, No.1, (June 2005), pp. 90–96, ISSN 0003-2670.

John, R; Wallace, G.G. 1990. The use of microelectrodes as substrates for chemically modified sensors A comparison with conventionally sized electrodes. *Journal of Electroanalytical Chemistry*, Vol. 283, No 1-2, (April 1990), pp. 87-98, ISSN 0022-0728.

Kelley, A; Angolia, B; Marawi, I. (2006).Electrocatalytic activity of poly(3-methylthiophene) electrodes. *Journal of Solid State Electrochemistry*, Vol.10, No.6, (June 2006), pp. 397-404, ISSN 1432-8488

Korri-Youssoufi, H; Garnier, F; Srivastava, P; Godillot, P; Yassar, A. (1997). Toward biolectronics: Specific DNA recognition based on an oligonucleotide-funcionalized pyrrole. *The Journal of American Society of Chemistry*, Vol.119, No.31, (August 1997), pp. 7388-7389, ISSN 0002-7863

Lange, U; Raznyatovskaya, N.V; Mirsky, V.M. (2008). Conducting polymers in Chemicals sensors and array. *Analytica Chemica Acta*, Vol.614, No.1, (April 2008), pp. 1-26. ISNN 0003-2670

Lupu, S; Parenti, F; Pigani, L; Seeber, R; Zanardi, C. (2003). Differential pulse techniques on modified conventional-size and microelectrodes. Electroactivity of Poly[4,4_-bis(butylsulfanyl)-2,2-bithiophene] coating towards dopamine and ascorbic acid oxidation, *Electroanalysis*, Vol.15, No.8, (June 2003), pp. 715-725, ISSN 0022-0728

Mark, Jr.H.B; Atta. N; Ma, Y.L; Petticrew, K.L; Zimmer, H; Shi,Y; Lunsford, S.K; Rubinson, F.J; Galal, A. (1995). The electrochemistry of neurotransmitters at conducting organic polymer electrodes: electrocatalysis and analytical aplications. *Bioelectrochemistry and Bioenergitics*, Vol.38, No.2, (October 1995), pp. 229-245, ISSN 032-4598.

McQuade, T.D; Pullen, A.E; Swager, T.M. (2000). Conjugated plymer-based Chemicals sensors. *Chemical Reviews*, Vol.100, No.7, (June 2000), pp. 2537-2574, ISSN 0009-2665

Mohan, S; Nigam, P; Kundu, S; Prakash, R. (2010). Label-free genosensor for BRCA1 related sequence based on impedance spectroscopy. *Analyst*, Vol.135, No.11, (November 2010), pp. 2887-2893, ISSN 0003-2654

Mousavi, Z; Bobacka, J; Ivaska, A. (2005). Potentiometric Ag^{+} Sensors Based on Conducting Polymers: A Comparison between Poly(3,4-ethylenedioxythiophene) and Polypyrrole Doped with Sulfonated Calixarenes. *Electroanalysis*, Vol.17, No.18, pp. 1609-1615, (September 2005), ISSN 0022-0728

Mousavi, Z; Bobacka, J; Lewenstam, A; Ivaska, A. (2006). Response mechanism of potentiometric Ag+ sensor based on poly(3,4-ethylenedioxythiophene) doped with silver hexabromocarborane. *Journal of Electroanalytical Chemistry*, Vol.593, No.1-2, (August 2006), pp. 219–226, ISSN 0022-0728

Muhammad-Tahir, Z; Alocilja, E.C. (2003). A conductometric biosensor for biosecurity. *Biosensors and Bioelectronics*, Vol.18, No.5-6, (August 2003), pp. 813-819, ISSN 0956-5663

Muthukumar, C; Kesarkar, S. D; Srivastava, D. N. (2007) Conductometric mercury [II] sensor based on polyaniline–cryptand-222 hybrid. *Journal of Electroanalytical Chemistry*, Vol.602, No.2, (April 2007), pp. 172–180, ISSN 0022-0728

Nambiar, S; Yeow, J.T.W. (2011). Conductive polymer-based sensor for biomedical application. *Biosensors and Bioelecronics*, Vol.26, No.5, (January 2011), pp. 1825-1832, ISSN 0956-5663

Izaoumen, N; Bouchta, D; Zejli, H; El Kaoutit, M; Stalcup, A.M; Khalid R. Temsamania, K.R. (2005). Electrosynthesis and analytical performances of functionalized poly(pyrrole/β-cyclodextrin) films. *Talanta*, Vol.66, No.1, (March 2002), pp. 111–117, ISSN 0039-9140

Pandey, P.C; Singh, G; Srivastava, P.K. (2002). Electrochemical synthesis of Tetraphenylborate doped Polypyrrole and its applications in designing a novel Zinc and Potassium ion sensor, *Electroanalysis*, Vol.14, No.6, (March 2002), pp. 427-432, ISSN0022-0728

Pardieu, E; Cheap, H; Vedrine, C; Lazerges, M; Lattach, Y; Garnier, F; Remita S; Pernelle, C. (2009). Molecularly imprinted conducting polymer based electrochemical sensor for detection of atrazine. *Analytica Chimica Acta*, Vol.649, No.2, (September 2009), pp.236–245, ISSN 0003-2670

Peng, H; Zhang, L; Soeller, C; Travas-Sejdic, J. (2009). Conducting polymers for electrochemical DNA sensing. *Biomaterials*, Vol.30, No.11, (April 2009), pp. 2132-2148. ISSN 0142-9612

Pickup, N.L; Shapiro, J. S; Wong, D.K.Y, (1998). Extraction of silver by polypyrrole films upon a base±acid treatment. *Analytica Chimica Acta*, Vol.364, No.1-3, (May 1998), pp. 41-51, ISSN 0003-2670

Prakash, R; Srivastava, R.C; Pandy, P.C. (2002). CopperII ion sensor based on electroplymerized undoped conducting polymers. *Journal of Solid State Electrochemistry*, Vol.6, No.3, (March 2002), pp. 203-208, ISSN 1432-8488

Rahman, M.A; Won, M-S; Shim, Y-B. (2003). Characterization of an EDTA bonded conducting polymer modified electrode: Its application for the simultaneous determination of heavy metal ions. *Analytical Chemistry*, Vol.75, No.5, (March 2003), pp. 1123-1129, ISSN 0003-2700

Rahman, Md.A; Kumar, P; Park, D-Su; Shim, Y-B. (2008). Electrochemical sensor based on organic conjugated polymers. *Sensors*, Vol.8, No.1, (January 2008), pp. 118-141, ISNN 1424-8220

Santhanan, K.S.V. (1998). Conducting polymers for biosensors: Rational based medels. *Pure & Apllied Chemistry*, Vol.70, No.6, (June 1998), pp. 1259-1262, ISSN: 0033-4545

Sargent, A; Loi, T; Susannah, G; Sadik, A.O. (1999). The electrochemistry of antibody-modified conducting polymers electrodes. *Journal of Electroanalytical Chemistry*, Vol.470, No.2, (July 1999), pp. 144-156, ISSN 0022-0728

Schalkhammer, T; Mann-Buxbaum, E; Pittner, F. (1991). Electrochemical glucose sensors on permselective non-conducting pyrrole polymers, *Sensors and Actuators B*, Vol. 4, No.3-4, (June 1991), pp. 273-281, ISSN 0925-4005

Schuhmann, W; Lammert, B.U; Schmidt, H.-L. (1990). Polypyrrole, a new possibility for covalent binding of oxidoreductases to electrode surfaces as a base for stable biosensors. *Sensors and Aciuator B*, Vol.1, No.1-6, (January 1990), pp. 537-541, ISSN 0925-4005

Schuhmann, W; Kranz, C; Huber, J; Wohlschläger, H. (1993). Conducting polymer-based amperometric enzyme electrodes. Towards the development of miniaturized reagentless biosensors. *Synthetic Metals*, Vol.61, No.1-2, (November 1993), pp. 31-35, ISSN 0379-6779

Song, F.Y; Shiu, K.K. (2001). Preconcentration and electroanalysis of silver species at polypyrrole film modified glassy carbon electrodes. *Journal of Electroanalytical Chemistry*, Vol.498, No.1-2, (February 2001), pp. 161-170, ISSN 0022-0728

Theâvenot, D.R; Toth, K; Durst, R.A; Wilson, G. (1999). Electrochemical biosensors: recommended definitions and classification. *Pure & Applied Chemistry*, Vol.71, No.12, (December 1999), pp. 2333-2348, ISSN 0033-4545

Thompson, B.C; Winther-Jensen, O; Vongsvivut, J; Winther-Jensen, B; MacFarlane, D.R. (2010). Conducting polymers enzymes alloys: electromaterials exhibiting direct electron transfer. *Macromolecular Rapid Communications*, Vol.31, No.14, (July 2010), pp. 1293-1297, ISSN 1022-1336

Vázquez, M; Bobacka, J; Luostarinen, M; Rissanen, K; Lewenstam, A; Ivaska, A. (2005a). Potentiometric sensors based on poly(3,4 ethylenedioxythiophene) (PEDOT) doped with sulfonated calix[4]arene and calix[4]resorcarenes. *Journal of Solide State Electrochemistry*, Vol.9, No.5, (May 2005) pp. 312-319, ISSN 1432-8488

Vázquez, M; Bobacka, J; Ivaska, A. (2005b). Potentiometric sensors for Ag+ based on poly(3-octylthiophene) (POT). *Journal of Solide State Electrochemistry*, Vol.9, No.12, (December 2005), pp. 865-873, ISSN 1432-8488

Vestergaard, M; Kerman, K; Tamiya, E. 2007. An Overview of Label-free Electrochemical Protein Sensors. *Sensors*, Vol.7, No.12, (December 2007), pp. 3442-3458, ISSN 1424-8220

Wang, J; Chen, I.S-P; Lin, M.S. (1989). Use of different electropolymerization conditions for controlling the size-exclusion selectivity at polyaniline, polypyrrole and polyphenol films. *Journal of Electroanalytical Chemistry*, Vol.273, No.1-2, (November 1989), pp. 231-242, ISSN 0022-0728

Wang, Z; Liu, S; Wu, P; Cai, C. (2009). Detection of glucose based on direct electron transfer reaction of glucose oxidase immobilized on highly ordered polyaniline nanotubes. *Analytical Chemistry*, Vol.81, No.4, (February 15 2009), pp.1638-1645, ISSN 0003-2700

Wolwacz, S.E; Smolander, M; Crompton, T; Lowe, C.R. (1992). Covalent electropolymerization of glucose oxidase in polypyrrole, *Analytical Chemistry*, Vol.64, No.14, (July 1992), pp. 1541-1545, ISSN 0003-2700

Xiao, Y; Li, C.M; Liu, Y. (2007). Electrochemical impedance characterization of antibody-antigen interaction with signal amplification based on polypyrrole–streptavidin. *Biosensors and Bioelectronics*, Vol.22, No.12, (June 2007), pp. 3161-3166, ISSN 0956-5663

Yon-Hin, B.F.Y; Smolander, M; Crompton, T; Lowe, C.R. (1993). Covalent electropolymerization of glucose oxidase in polypyrrole. Evaluation of methods of pyrrole attachment to glucose oxidase on the performance of electropolymerized glucose sensors, *Analytical Chemistry*, Vol.65, No.15, (August 1993), pp. 2067–2071, ISSN 0003-2700

Zhang, H; Lunsford, S.K; Marawi, I; Rubinson, J.F; Mark, Jr.H.B (1997).Optimization of preparation of poly(3-methylthiophene)-modified Pt microelectrodes for detection of catecholamines. *Journal of Electroanalytical Chemistry,* Vol.424, No.1-2, (March 1997), pp. 101- 111, ISSN 0022-0728

Zhao, M; Wu, X; Cai, C. (2009). Polyaniline nanofibers: synthesis, characterization, and application to direct electron transfer of glucose oxidase. *The Journal of physical Chemistry C,* Vol.113, No.12, (March 2007) pp. 4987-4996. ISSN 1932-7447

Zanganeh,A.R; Amini, M.K. (2007) A potentiometric and voltammetric sensor based on polypyrrole film with electrochemically induced recognition sites for detection of silver ion . *Electrochimica Acta,* Vol. 52, No. 11, (March 2007), pp. 3822–3830, ISSN 0013-4686

Zejli, H; Izaoumen, N; Bouchta, D; El Kaoutit, M; Temsamani, K.R. (2004). Electrochemically aided solid phase micro-extraction of mercury(II) at poly(3-methylthiophene) modified gold electrode. *Analytical Letters,* Vol.37, No.8, (Not found), pp. 1737-1754, ISSN 0003-2719

4

Ionophore/Lipid Bilayer Assembly on Soft Organic Electrodes for Potentiometric Detection of K^+ Ions

Osvaldo Abreu, Jeannine Larrieux and Kalle Levon
Department of Chemical and Biological Sciences
Polytechnic Institute of NYU
Six Metrotech Center, Brooklyn, New York
USA

1. Introduction

The major goal on intrinsically conducting polymers (ICP) technology development has been to combine the electrical and optical properties of these new materials with the mechanical and processibility properties of commodity bulk polymers. New conductive materials that offer significant application potential as substitutes, and new products having properties difficult or impossible to achieve by existing materials, can now be produced.

The most common synthetic methods are the oxidative chemical polymerization of aniline using ammonium persulfate, electrochemical polymerization; and most recently enzyme catalyzed polymerization of aniline has also been reported (Jin 2001).The reactions of aniline with oxidants proceed in two fundamentally different ways. The oxidant can either donate oxygen to the aniline molecule or it can remove a hydrogen atom from its amino group, the latter is the prevalent method that leads to polyaniline (PANI) (Ćirić-Marjanović 2008). It was shown by Wan and co-workers that during oxidative chemical polymerization, the molar ratio of acid to aniline affected the probability of formation of fibrous nanostructures dictating the final product morphology (Zhang 2002).

In the oxidative chemical polymerization method aniline and ammonium persulfate are mixed at a predetermined stoichiometric ratio and reacted in acidic medium producing a green precipitate of PANI salt. For the electrochemical method an applied potential in lieu of an oxidant is used to oxidize and polymerize the aniline on the anodic electrode surface. In 2002 the International Union of Pure and Applied Chemistry (IUPAC) in an effort to standardize the process of oxidative polymerization of aniline using ammonium persulfate, issued a Technical Report with the goal of obtaining PANI with a defined conductivity (Stejskal 2002). This report constituted the first organized robustness study of the oxidative chemical polymerization of aniline, and it included eight chemists from five different institutions executing the same IUPAC preparation protocol. The electrochemical method was excluded from this report since its efficiency has been proved to be a function of the electrode area and therefore not suitable for large-scale production. An efficient polymerization of aniline is achieved only in an acidic medium where aniline exists as the anilinium cation. The reaction is exothermic with a defined and reproducible temperature profile that can be used to monitor the progress of the reaction (Beadle 1998). Generally there is an induction period of about 4 minutes where the temperature stays constant at pH ≤ 2.5, followed by polymerization period where the temperature raises close to 40°C, then the temperature start falling during the post polymerization period.

The presence of residual aniline must be minimized to obtain the best yield of PANI and a stoichiometric ratio peroxydisulfate/aniline of 1.25 is recommended, the polymerization is complete within 10 minutes at room temperature, however it slows down to 1 hour when the reaction container is cooled between 0-2 °C (Sulimenko 2001). PANI has a nitrogen heteroatom incorporated between the phenyl rings along the polymer chain imparting flexibility and allowing the existence of the six different forms known (Naarman 1987) as seen in Figure 1 and Figure 2. The PANI powder synthesized by the chemical oxidative polymerization in acidic solution is the emeraldine salt (ES) form which when treated with alkali becomes the emeraldine base (EB) form. Oxidation and reduction of the ES form yields the pernigraniline salt (PNS) form and the leucoemeraldine salt (LES) form respectively, while same treatment of the EB form produces the pernigraniline base (PNB) and the leucoemeraldine base (LEB). The individual base and salt forms can convert into each other by protonation/deprotonation.

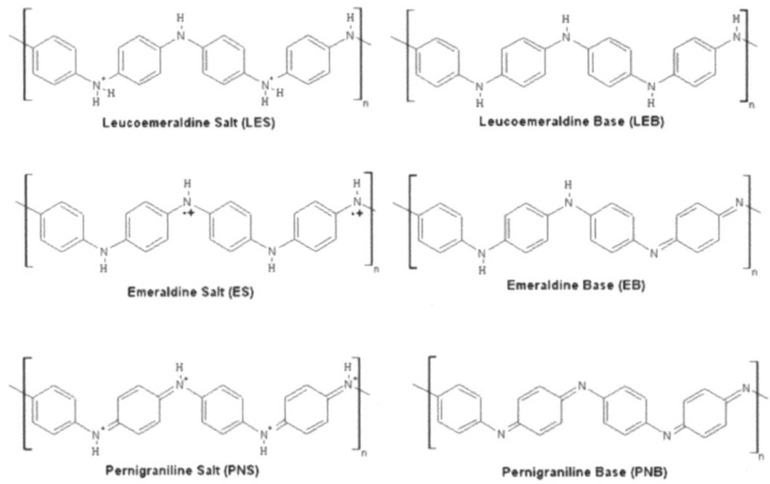

Fig. 1. Chemical structures of the six different polyaniline chemical forms

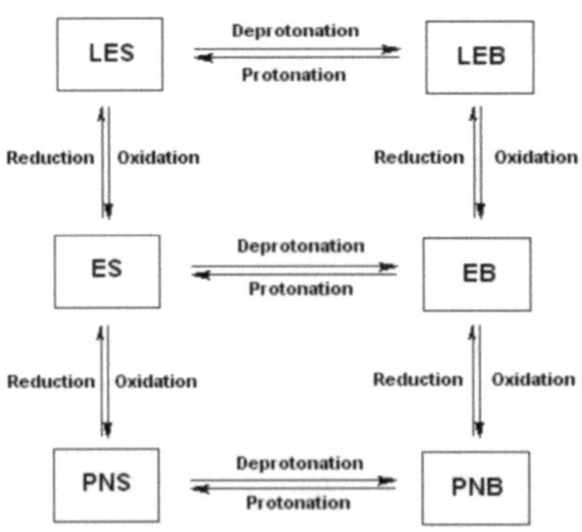

Fig. 2. Polyaniline interconversion chart for the six different chemical forms

Polyaniline has a conduction mechanism that is unique among conductive polymers in that its most highly conducting doped form can be reached by two different processes:

- Protonic acid doping of the EB form and
- Oxidative p-type doping of the LEB form.

The term doping was coined to refer to the introduction of impurities in an extremely pure semiconductor in order to change its electrical properties. However this terminology has also been used to refer to a process consisting in the introduction of charge carriers in ICPs. Charge carriers are created by removing or adding electrons to the delocalized π-electrons through a mechanism that is more precisely termed as a redox reaction (Zhang 2001).

Through the process of doping the number of electrons in the PANI chains are either decreased, or increased resulting in structural defects in the bond interaction and localized electronic states. Since the geometry of the chain between the radical and the ion must be distorted, and the energy of the distorted geometry is normally higher than that of the ground state, separation, and delocalization of the radical ion associated with a lattice distortion are called polarons and are the basis of the hopping conductivity typical of ICPs.

2. Functionalization of polyaniline

Several solutions to the intractability of PANI have been devised throughout the years, all these solutions aim to improve solubility and processability in organic solvents. Large organic anions such as dodecylbenzenesulfonate and camphorsulfonate are used as dopants, also sulfonation has also been used to improve solubility and processability in aqueous media. Zheng investigated the incorporation of flexible alkyl chains into polyaniline through N-alkylation of PANI LEB form to improve it tractability, and found that long alkyl side chains facilitate the formation of mesophases by functioning as flexible spacers to decouple individual mesogenic units for the entire polymer chain (Zheng 1994). Han et al through review and observation that many Lewis base solvents, e.g. pyrrolidine, piperidine and morpholine are effective solvents for the PANI EB form, proved that the class of compound provided concurrent reduction and substitution between PANI and nucleophiles, and therefore a facile approach for modifying a PANI backbone with various electron-donating groups, e.g. amino, alkylthio and alkoxy (Han 1997). The protonation of the imine nitrogen could possibly promote nucleophilic attacks at the meta position of the protonated quinoid ring, followed by a 1,3 proton shift of the meta-proton to the para-imine nitrogen and, ultimately, a conversion of the less stable quinoid ring to the more stable benzenoid. Highly conductive aniline copolymers containing butylthio substituents have been prepared from unsubstituted polyaniline confirming that a concurrent reduction and substitution route performed on the solid-state matrix of PANI is a better way for preparing aniline copolymers, than the conventional copolymerization (Han 2001). The butylthioaniline copolymers obtained by this process are highly soluble in common organic solvents, such as THF, dioxane, 2-methoxyethyl ether, and 2-methoxyethanol [47] which are nonsolvents for the parent PANI.

3. Electrochemical properties of polyaniline

The most useful electroanalytical technique for the characterization of PANI is cyclic voltammetry (CV), in this technique the potential applied across the electrode-solution interface is varied and the resulting current is measured. The versatility of CV allows a mechanistic study of redox systems, this is of great utility for the identification of peaks associated with the PANI redox couples, further characterization can be done using the peaks individual potentials in the voltammogram, and by changing the scan rate.

A typical cyclic voltammogram of PANI in aqueous 0.5N sulfuric acid is shown on Figure 3, two well resolved peaks at half-potential ca. 0.2 and 0.7 V corresponding to two quasi-reversible redox processes, the first corresponds to the conversion of PANI protonated amine units (LES) to the semiquinone radical cations in ES, while the the second can be assigned to the conversion of ES to the fully oxidized and protonated quinoidal form PNS.

The shape of the cyclic voltammogram is independent of the nature of the aqueous acid electrolyte only when strong acids are used (Batich 1990), however, the redox potentials strongly depend on the film thickness, scan rate, and pH of the medium. As the pH is increased the redox couple peaks shift to a higher positive potential, this is due to differences between partial and complete protonations of the amine nitrogens (Lukachova 2003, Varela 1990). The composition of the electrolyte medium has also been found to influence the redox mechanism, since both the cation and anion present participate in the PANI redox mechanism Genies 1990). Electrogravimetric studies in aqueous media (DeSilvestra 1992, Kabumoto 1988) have shown that the mass of PANI polymer increases during oxidation (charging) and decreases during reduction (discharging) for the first redox couple. Mass changes depend upon the molecular weight of the anion, and there is not effect from the cation, this is an indication that the anions present in solution are combined into the polymer films to compensate for the positive charges resulting from the oxidation from LEB to ES, and then driven out during subsequent reduction.

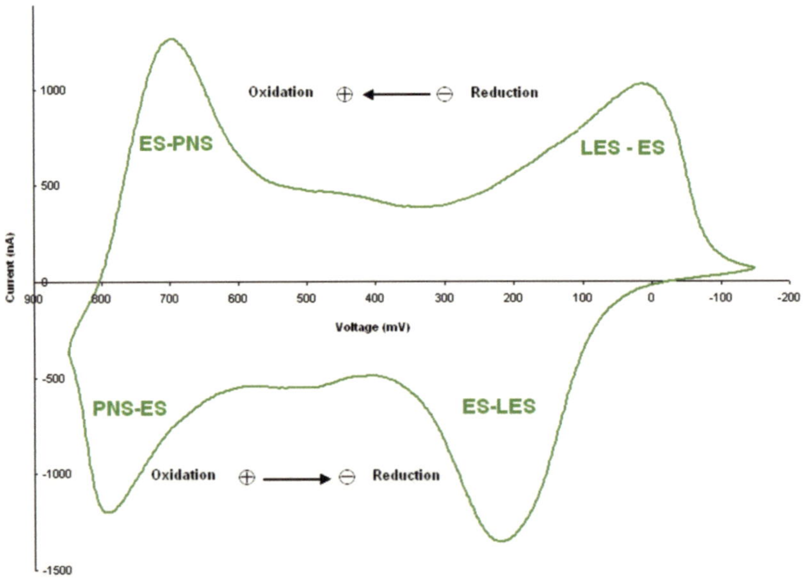

Fig. 3. Cyclic voltammogram of a polyaniline film on glassy carbon microelectrode (7 mm^2 disk in 0.5M H_2SO_4 from -0.150V to 0.850 V, scan rate = 100 mV/s.

4. Ion-Selective polyaniline electrodes

Today, conducting polymers have become one of the most important ion-to-electron transducers in solid-state ion selective electrodes (ISEs). They are based on the potentiometric measurement technique which is very attractive for practical applications because it allows the use of small-size, portable, low-energy consumption and relatively low-cost instrumentation[57]. Furthermore, the development of solid-state ISEs without internal filling solution gives durable and maintenance-free ion sensors. The principle of ion-to-electron transduction is derived from the fact that ICPs are polymers with conjugated double bonds, and thus considered electroactive materials that have widespread use in the field of chemical sensors. Oxidation of the conjugated polymer backbone is accompanied by anion insertion or cation expulsion, as shown on equation 1 and 2.

$$P + A^- \rightleftarrows P^+A^- + e^- \quad (1)$$

$$PA^-M^+ \rightleftarrows P^+A^- + M^+ + e^- \quad (2)$$

Where P = neutral conducting polymer unit, P^+ = oxidized conducting polymer unit, A^- = anion, M^+ = cation and e^- = electron. Equations 1 and 2 describe two limiting cases where either anions or cations are mobile, respectively. The similarity between these two equations immediately suggests that conducting polymers can work as ion-to-electron transducers in ISEs as they are frequently used today.

5. Polymer supported biomimetic membranes

Lipid-bilayer membranes supported on solid substrates are widely used as cell-surface models that connect biological and artificial materials (Turner 1996). They can be placed either directly on solids or on ultrathin polymer supports that mimic the generic role of the extracellular matrix. Biological membranes consist largely of a lipid bilayer that imparts a fluid character, there are proteins embedded in it and carbohydrates attached to its surface which facilitates communication and transport across it.

The complexity of biological membranes and their interactions with intra and extracellular networks make direct investigation difficult and for this reason artificial membranes have played an important part in unraveling the physical and chemical characteristics of membranes and their contributions to membrane function. The most commonly used experimental cell-surface model for the past 20 years have consisted of phospholipid bilayers deposited onto solid substrate. Solid-supported membranes are prepared by directly depositing lipid monolayers and bilayers onto surfaces, which maintain excellent mechanical stability without losing their fluid nature (Tanaka 2005, Groves 2003 and Sackmann 1996).

The fluidity and stability of the solid-supported membranes planar surfaces have a distinct advantage over freestanding "black lipid" membranes or spherical lipid vesicles models because it makes it possible to carry out experiments and use analytical methods that are difficult or impossible to use with other model systems. Methods like total interference fluorescence, nuclear magnetic resonance (NMR), Fourier-transform infrared spectroscopy, Surface Plasmon resonance, and X-ray and neutron scattering can all be used to probe the structural and dynamic properties of solid supported membranes (Kalb 1992, Baruerl 1990, Tatulian 1995, Terretaz 1993, Kjaer 1987, Kalb 1990, Majeski 1998).

In solid-supported membranes the artificial membranes and their solid supports are close together. They typically approach each other to within 5-20Å. This small gap leaves a water reservoir that is usually not sufficient to prevent protein subunits from coming into direct contact with the bare substrate. Such direct contact can be avoided by using polymer supports of typically less than 100 nm thickness that "cushion" or "tether" the membrane.

When using a polymer to 'cushion' a supported membrane, it usually acts as a lubricating layer between the membrane and the substrate. This will assist self-healing of local defects in the membrane over macroscopically large substrates (~cm2) and allow the incorporation of large scale transmembrane proteins without the risk of direct contact between protein subunits and the bare substrate surface. J. Majewski and J.N. Israelachvili prepared softly supported polymer-cushioned membranes which consisted of thin layer of branched, cationic polyethyleneimine (PEI), and the bilayers were formed by adsorption of small unilamellar dimyristoylphosphatidylcholine (DMPC) vesicles (Wong 1999). Recently glycolipid-containing bilayer have been shown to be an effective tethering system (Lipkowski 2010, Chen 2009, Brosseau 2008).

Our objective of this research is the derivatization of PANI with thiolated phospholipids, with the purpose of assembling covalently bonded phospholipid layers on this ICP. The EB form was the starting material, as its chemical structure presents one unique opportunity for derivatization though a concurrent substitution and reduction of its diiminoquinoid rings to diaminobenzenoid rings forming the PANI leucoemeraldine base (LEB) form. Alkylthiols and alkylamine nucleophiles have been used to derivatize EB, obtaining the LEB reduction product with an assembled hydrophobic layer on it; in a similar fashion as thiols self-assemble on gold surfaces. However as far as we know, this work has not yet been attempted with thiolated phospholipids. As part of this research some thiolated phospholipids were also synthesized by treating 1,2-dipalmitoyl-sn-glycero-3-phosphoethanolamine (DPPE) with heterobifunctional cross-linkers also known as thiolation agents, which are commonly used for coupling $-NH_2$ containing compounds and –SH containing molecules together by a covalent bond. However this work required multiple synthesis/separation steps and exhaustive purification which were suspended once a thiolated phospholipid commercial product named 1,2-dipalmitoyl-sn-glycero-3-phosphothioethanol sodium salt (PTE) was found to be available.

Once PTE was substituted on the PANI films supported on carbon glassy (CG) electrodes, attempts were made to assemble K^+ ion specific electrodes, based on the immobilization of the ionophore valinomycin and the pentadecapeptide gramicidin forming biomimetic lipid membranes. Potentiometric titrations with K^+ ions were performed to determine the assembled membranes saturation curves, and to determine if the mechanism of adsorption was consistent common cellular facilitated transport mode.

Phospholipid, also known as phosphoglycerides, are triglyceride-derivatives in which one fatty acid has been replaced by a phosphate group and one of several nitrogen-containing molecules. The hydrocarbon chains are hydrophobic as in all fats. However, the charges on the phosphate and amino groups make that portion of the molecule hydrophilic resulting in an amphiphilic molecule.

As amphiphilic molecules, the phospholipids are major constituents of the cell membrane wherein with their hydrophilic polar heads face aqueous surroundings like the cytosol and their hydrophobic non-polar tails face each other. If suspended in water, these molecules can orient themselves at the air/water interface with their hydrophobic chains upwards in the air and their hydrophilic head groups downwards into the aqueous phase. The chemical structures of 1,2-dipalmitoyl-sn-glycero-3- phosphothioethanol sodium salt (PTE) and 1,2-Dipalmitoyl-sn-Glycero-3-Phosphoethanolamine (DPPE) are shown in Figure 4 as A and B, respectively. The preparation of thiolated phospholipids requires a primary amine group as observed in DPPE.

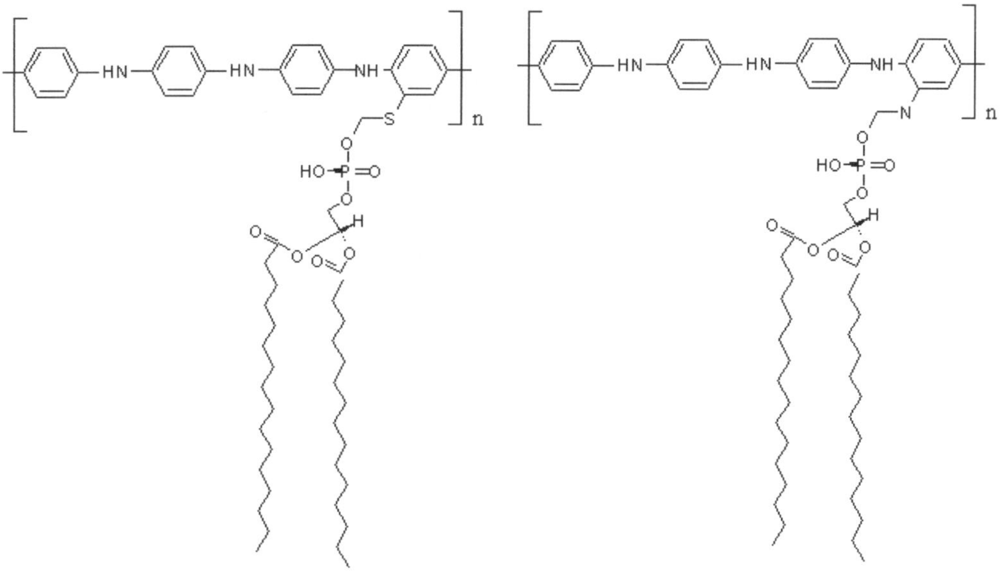

Fig. 4. 1,2-dipalmitoyl-sn-glycero-3- phosphothioethanol sodium salt (PTE) and 1,2-Dipalmitoyl-sn-Glycero-3-Phosphoethanolamine (DPPE)

6. Adsorption of amphiphilic molecules at electrode interfaces

A phenomenon that can markedly affect the results of electrochemical experiments, is the tendency of amphiphilic ions or molecules to adsorb on the electrode surface and their aggregation into supramolecular structures (Rusling 1991). Neutral molecules can adsorb on the electrode surface primarily because of the hydrophobicity in aqueous solution. Generally, the less soluble a molecule is, the stronger it adsorbs (Kissinger 1996). Bonding between the electrode surface and π electrons and nitrogen lone pairs electrons (as in -SH and -NH_2) of the molecule can also enhance adsorption.

The PANI coated Pt and GC electrodes used in this research has proven to be a more versatile approach to prepare a chemically modified electrode (CME) that just a simple polymer coat as initially shown by Miller and Bard (Miller 1978, Merz 1978). The diiminoquinoid rings in the PANI film backbone can form a covalently bonded densely packed layer of phospholipids, primarily because of increased hydrophobicity of the electrodes which facilitates the approach of the phospholipid and the eventual functionalization of the backbone by the -SH group.

When redox reagents are absorbed on an electrode surface or confined on a thin layer of solution adjacent to it, the reversible cyclic voltammograms are not governed by diffusion controlled processes and this phenomenon can markedly affect the results of any electrochemical experiment. Adsorption is responsible for unusual electrochemical behavior[2] and for this research it played an important role in enhancing the substitution of PTE on PANI.

Derivatization of electrochemically synthesized PANI films on GC electrode was performed by using DPPE phospholipid which contains $-NH_2$ nucleophile. The chemical structure of DPPE is shown on Figure 4. DPPE vesicles were made by followingly: The procedure consisted of weighing 5.2 mg of DPPE into a glass scintillation vial, dissolved in 2 mL of chloroform, then evaporated under a stream of nitrogen, and dried under vacuum for 60 minutes. The DPPE solid was then hydrated in 8 mL of 0.5M H_2SO_4 by intermittent agitation within a period of 60 minutes (shaking every 10 minutes), then the mixture was frozen and thawed five times using a salt-water ice bath and a water bath at 37°C, and extruded by filtering through an Acrodisc GHP 0.45μm syringe filter after each freeze/thaw cycle . The final solution contained DPPE small unilamellar vesicles (SUVs), which were transferred to an electrochemical cell. A PANI coated GC electrode was used as the working electrode, a platinum wire as the counter electrode, and an Ag/AgCl as reference electrode. The cell was cycled 50 times from -0.150V to +0.850V at scan rate of 100 mV/s. For comparison, the same procedure was repeated to prepare PTE SUVs which were then used, to treat an additional PANI GC electrode using the same procedure described above

CVs of conventional PANI in acidic solutions show two redox couples[11], as shown in Figure 5, the first pair of peaks located at low potentials correspond to the LE-ES/ES-LES redox couples, while the second couple located at higher potential is due to the ES-PNS/PNS-ES. A review of the data show that the LES-ES transition peaks corresponding to the fully benzenoid radical cation, and the ES-LES transition both experimented a blue shift. Contrarily the ES-PNS transition peak corresponding to the fully oxidized quinoidal form showed a red shift while the PNS-ES transition peak disappeared after an initial red shift. Overlay CVs are shown on Figure 6.

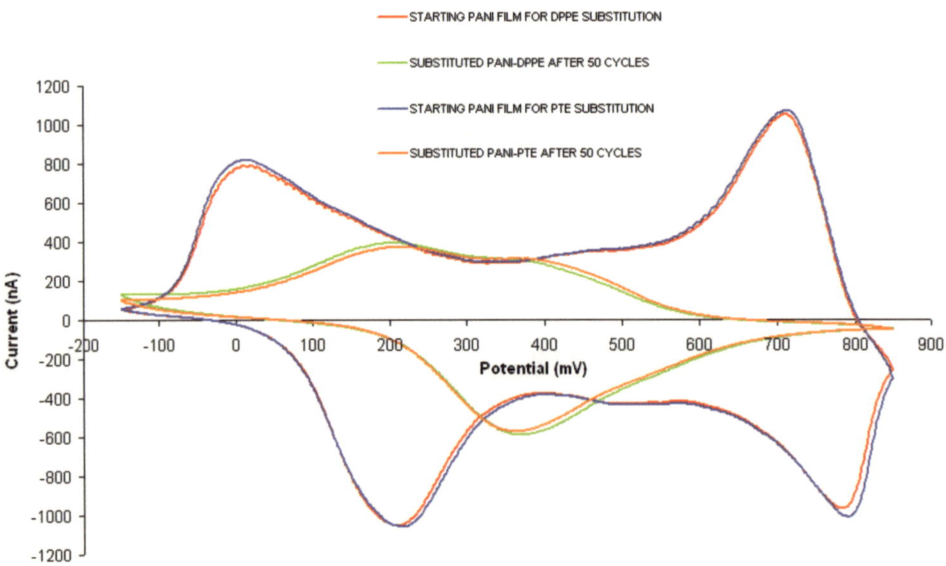

Fig. 5. Overlaid voltammograms for PANI films before and after the substitution with DPPE and PTE

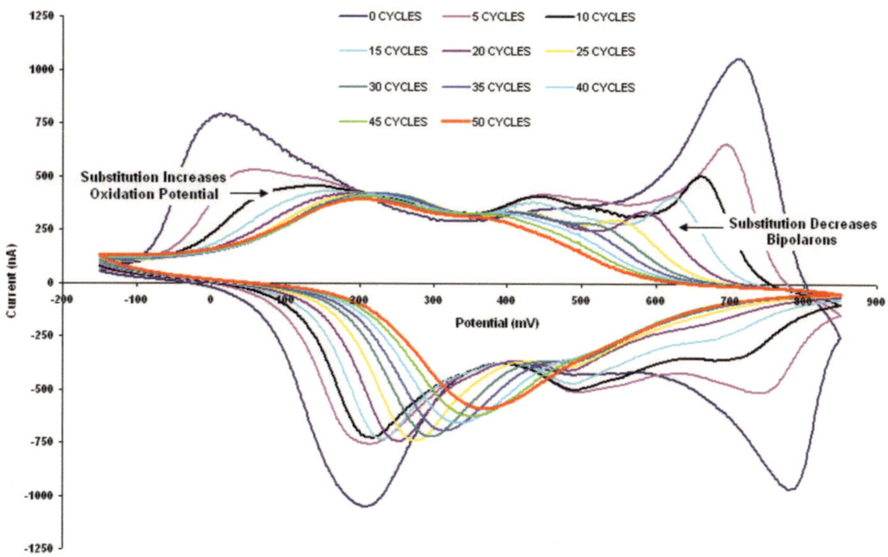

Fig. 6. Potentiodynamic substitution of PANI film with PTE

The overlaid evolution of the potentiodynamic substitution of PTE on a PANI film is displayed on Figure 6, it basically shows that the PNS-ES peak disappears after only 10 cycles, and then two new peaks on the anodic and cathodic planes appeared on the CV. It is believed that after the 10th cycle, any ES that is created by the potential cycling, thus reacted immediately with PTE to become the newly substituted LES product. The LES-ES shifts towards the center of the CV along with the remainder ES-PNS Peak which is continuously depleted by its transformation to ES and then to LES. At the end of the 50th cycle, only three peaks remain, the fully reduced and PTE substituted LES-ES transition, its corresponding ES-PNS radical cation and the blue shifted PNS-ES transition.

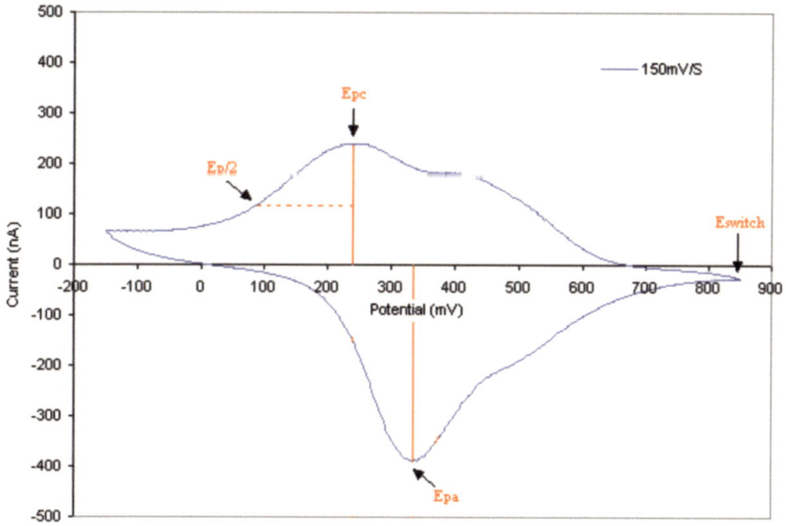

Fig. 7. Cyclic Voltammogram of PTE substituted PANI GC Electrode

The disk shaped GC electrode coated with derivatized PANI-PTE products showed a distinctive voltammogram, consistent with a gradual increase of the oxidation potential for the benzenoid forms and a decrease in redox potential for the diiminoquinoid forms, thus indicating a decrease on the bipolarons. The CV of the PTE substituted PANI is shown on Figure 7 using a scan rate of 150 mV/s

The CV on Figure 6 shows a gradual evolution of the substitution of PTE on the PANI film supported on the GC Electrode. At the end of 50th cycle one could wonder what could be the extent of the surface functionalization and how to measure it. Several techniques come to mind to make such measurements, some rather macroscopic like contact angle measurement could in fact measure the change in hydrophobicity on the PANI surface, other like ATRIR could measure the degree of substitution on the PANI film, and the use of a Crystal Quartz Microbalance (QCM) being an ultrasensity mass detector could measure the change in mass during the reaction progress. Finally electrochemical impedance spectroscopy (EIS) could also be used to measure the film capacitance. Film capacitance changes when lipids films attach to the electrode surface, it has been shown that lipid SAMS on gold electrodes can be modeled as flat plate capacitors in series (Rose 2006).

Due to the intrinsic geometry of the GC electrode, the GC disk is attached right at the bottom of a PTFE body. This makes it difficult to handle for placing on a conventional contact angle measurement apparatus. Same issue occurs with the ATRIR with requires a measurement tip to touch a horizontal section of the sample. To get a ATRIR spectrum the sample must fit under the ZnSe measurement tip and that could not be done with the GC electrode.

The synthesis of PANI on Pt and GC electrodes along with concurrent derivatization reactions, lead to the chemical and electrochemical substitution/bonding of phospholipids lipid layers on the conductive polymer substrate. The goal is to present the steps taken to assemble lipid membranes containing ion carrier proteins on electrodes coated with different conductive polymers, with aim of creating simulated natural systems. Once the membranes were assembled, their integrity and response was investigated by measuring their potentiometric responses by titration with K^+ ions.

7. Potentiometric detection

Potentiometry is one of the most frequently used analytical methods in chemical analysis[30]. This analytical technique is characterized by the measurement of the potential difference between two electrodes while maintaining the electric current under a nearly zero-current condition. In most common forms of potentiometry, the potential of the WE electrode varies depending on the concentration of the analyte, while the potential arising from a second RE electrode is ideally a constant. In practice the general principle of potentiometry deals with the electrochemical force (EMF) generated in a galvanic cell where a spontaneous chemical reaction takes place. The measurement of the electrochemical cell potential under zero-current conditions is used to determine the concentration of analytes in measuring samples.

PANI films have been used for the fabrication of ultramicroelectrodes sensitive to pH, showing some advantages over conventional glass electrodes (Slim 2008, Zhang, X 2002). Many other applications of PANI included potentiometric sensing of creatinine (Pandley 2004), dodecyl benzene sulfonate ISE (Karami 2004), nitrates ISE (Mazeikiene 1997), and quantitation of salinity (Diniz 1999). Films of other conductive polymers like PPy, poly(3,4-

ethylenedioxythiophene), and poly(3-octylthiophene) have been casted on solid supports to make potentiometric nanoelectrodes (Bakker 2008). PPy and PANI are considered today as the most promising conductive polymers for the development of sensors devices, owing to its good biocompatibility, conductivity, and stability.

The electrodes assembled during the experimental work fit in the category of all-solid state, as there was no need of using an internal reference solution. To avoid the drift problems observed on coated-wire electrodes (CWE), Cadogan et al. used an ICP as an intermediate layer between the ion selective membrane (ISM) and the substrate (Cadogan 1992), where the ICP was acting as an ion-to-electron transducer, thus eliminating the mismatch between the ionic conductivity in the membrane and the electronic conductivity in the substrate. This mismatch causes a high charge transfer resistance at the interface and the drift.

A drawing representation of the electrochemically derivatized PANI film is shown on Figure 8. The surface coverage was tested and showed that the thiolated phospholipid derivatized surface was hydrophophic, as indicated by the drift to negative potential. This also proved that the PANI-PTE surface did not exhibit ion exchange capabilities. At this point the surface became a platform for biomolecule immobilization as PANI could provide rapid electron transfer for use as a biosensor.

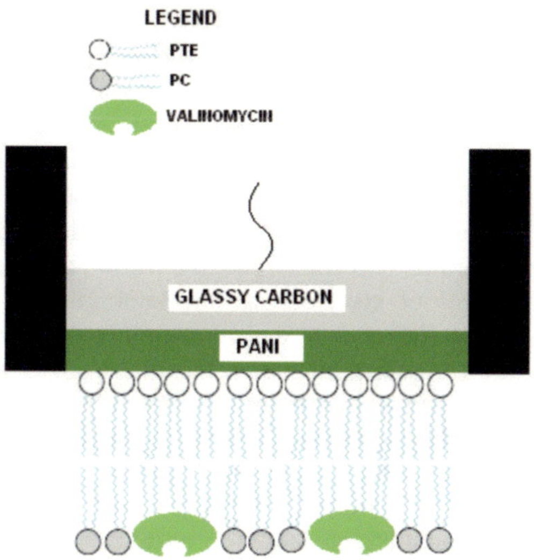

Fig. 8. Assembled bilayers with ionophore on GC electrode

In order to induce ion selectivity, ion-recognition sites as provided by ionophores must be immobilized in the derivatized PANI film, either covalently or non-covalently (Vasquez 2005). In solid contact as it was the case of the experiments performed on this research, PANI acts solely as an ion-to-electron transducer between the ISM and the electronically conducting substrate.

The cell membrane is lined with narrow protein-lined pores, known as ion channel proteins; these proteins are molecular entities that catalyze the flux of ions across bilayer lipid membranes. They exhibit three essential properties: first; an ability to catalyze high transport rates: second, an ability to select between closely related ions, e.g. Na^+ and K^+, and third, regulation by external stimuli such as ligand binding or a change in transmembrane voltage, that is the property of being gated. A classification of natural ion channels according to the most common gating stimulus are: change in membrane potential (voltage-gated), drugs or chemical transmitters (ligand-gated), and mechanical deformation (stretch-gated). Gating is thought to involve conformational changes of the ion channel which alters selective permeability. We applied gramicidin and valinomycin as the ionophores in this work.

Gramicidin is an antibiotic obtained form the bacterial species Bacillus Brevis, it is a linear pentadecapeptide usually called, gramicidin D, which is a heterogeneous mixture of six antibiotic compounds called Gramicidins A, B, and C at the levels of 80%, 6% and 14% respectively. Gramicidin has been used in the fabrication of ion-channel based biosensors [3], because this hydrophobic linear polypeptide forms channels in phospholipid membranes that are specific for monovalent cations.

Valinomycin is a potassium selective ionophore obtained from the cells of several Streptomyces strains, capable of increasing the transport of potassium across cell membranes and thereby causing damage to bacteria cells.

Eisenberg[41] considered the behavior of ion channels as enzymes, pointing out that ion channels modify the flux of ions the same way enzymes modify the flux of reactants, by stabilizing the transition state between substrate and product. Ion channels also have substrate and products just as clearly defined as do enzymes. The substrate of a channel is just the permeable ion on one side of the membrane, and the product is just the permeable ion on the other side of the membrane. The transition state of one ion-channel consist basically of a step when the ion is in the pore, there the protein provides polarization charge to neutralize the permanent charge of the permeating ion. To continue describing the analogy, the substrate and product before passing through an ion channel, have different free energies, just as they do for enzymes. Substrates and products of enzymes are different chemical species with different free energies at the same location. Substrates before and after passing through an ion channel (product) are the same chemical species, at different locations. The spatial gradient of electrochemical potential drives diffusion just as the chemical gradient free energy drives a chemical reaction. Therefore an ion channel is a catalyst for diffusion through membranes and a channel is an enzyme, even if it does not facilitate ordinary chemical reactions. Naumann and Knoll have developed an ion channel – lipid bilayer assembly using His-tag connection to the electrode offering the possibility to follow electroactive membrane protein functions (Nauman 2003, Ataka 2004). Lipkowsky have used glycolipid conjugation tethering the bilayer with the electrode (Lipkowski 2010, Brosseau 2008). The cushion acts effectively as a soft support for the bilayer membrane allowing its dynamic function. The only example of assembling valinomycin ion channel on a conducting polymer-based organic electrode has been published by Bobacka et al who physically mixed valinomycin as the ionophore with poly(octyl thiophene) and used the system effectively as a selective potassium potentiometer. We have prepared the ion channel within the bilayer and then deposit the system on a conductive polymer, polyaniline.

Thiolated PANI GC electrodes prepared were coated at the physiological temperature of 37°C, with large unilamellar vesicles (LUVs) mixed with the ion carrier protein valinomycin (mw 1111.36 g mol^{-1}). Proteins have the advantage of being amphiphilic and thus anchor within a membrane and float within it. The procedure is shown below:

Pipetted 2 mL of egg L-α-phosphatidylcholine (concentration of 20 mg/mL in chloroform) to glass scintillation vials and evaporated under a stream of nitrogen, then kept under vacuum for 1-hour. Added 1.3 mg of valinomycin , followed by 20 mL of 0.5M H_2SO_4 and vortexed for 5 minutes to make a suspension. Transferred an aliquot of the suspension onto a 10 x 75 mm glass centrifuge tube and carefully suspended one thiolated PANI GC electrode inside the tube, with the aid of Parafilm®, such that it did not touch the bottom of the tube. Placed in a temperature controlled bath for 60 minutes at 37°C. Removed gently from the tube, and slowly washed the PTFE electrode body with Milli-Q water, never touching the bottom of the electrode. Special care was taken to protect the assembled membrane by suspending the electrode on the electrochemical cell lid and equilibrating in 10 mL of Milli-Q water until a steady potential reading was reached.

8. Potentiometric titration using valinomycin PANI modified electrode

A glass electrochemical cell containing 10 mL of Milli-Q water was outfitted with a three holed PTFE lid. The valinomycin modified electrode, and the Ag/AgCl reference electrode used two holes, while the third hole was left open to add a KCl titrant solution. The two electrodes were connected with a BNC cable to a pHmeter operating in potential mode, and allowed to equilibrate until reaching a steady absolute potential reading (approximately 60 minutes). The pHmeter display was adjusted to relative potential mode, displaying a zero value before starting the addition of titrant. The titrant potassium chloride solution was prepared as follows:

Weighed 105.6 mg of potassium chloride and transferred to a 50 mL volumetric flask, dissolved and diluted to volume with Milli-Q water for a final concentration of 2.112 mg/mL. The titration results are summarized on Appendix C, the absolute potential (ABS) at equilibrium was -268.5mV. Measurements were taken at 2 minutes intervals between additions, stirring with a magnetic bar for the first 30 seconds, then stopping and taking the reading after the remainder 1.5 minutes. This time is usually enough for the potentiometer to display a stable reading. The titration curve is shown on Figure 9 and its characteristic Nernstian plot on Figure 10.

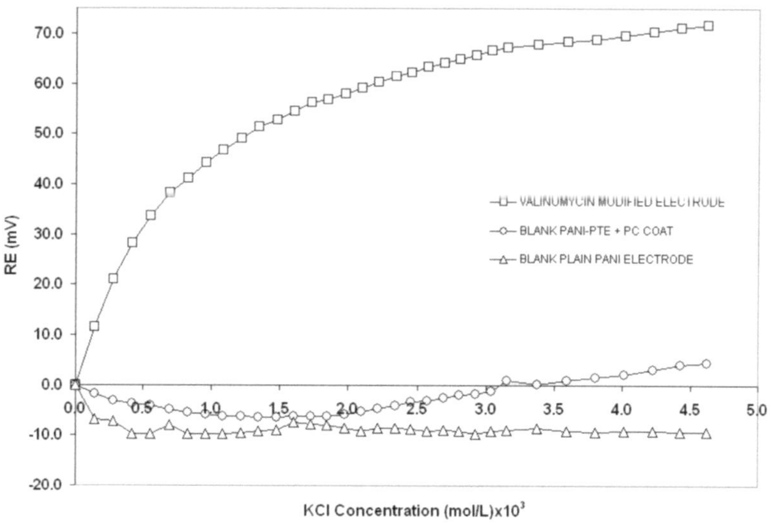

Fig. 9. Overlaid potentiometric titration curves for PANI-valinomycin modified electrode

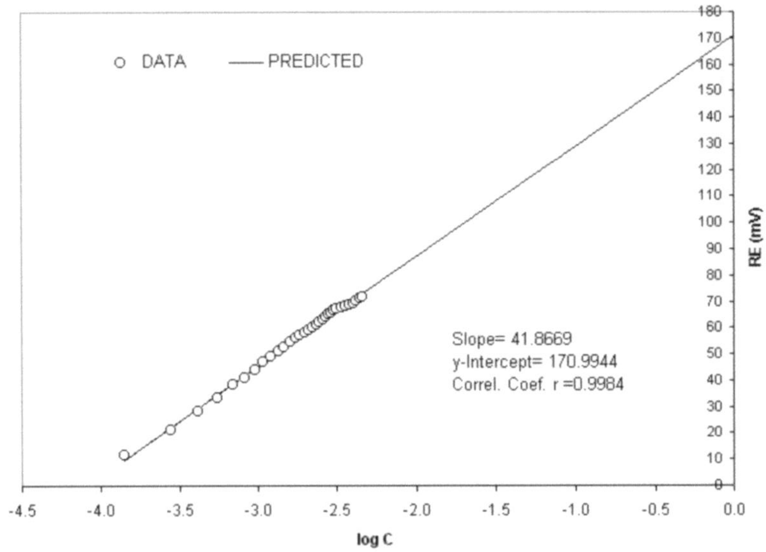

Fig. 10. Nernstian Plot for potentiometric Titration with PANI-valinomycin Modified Electrode

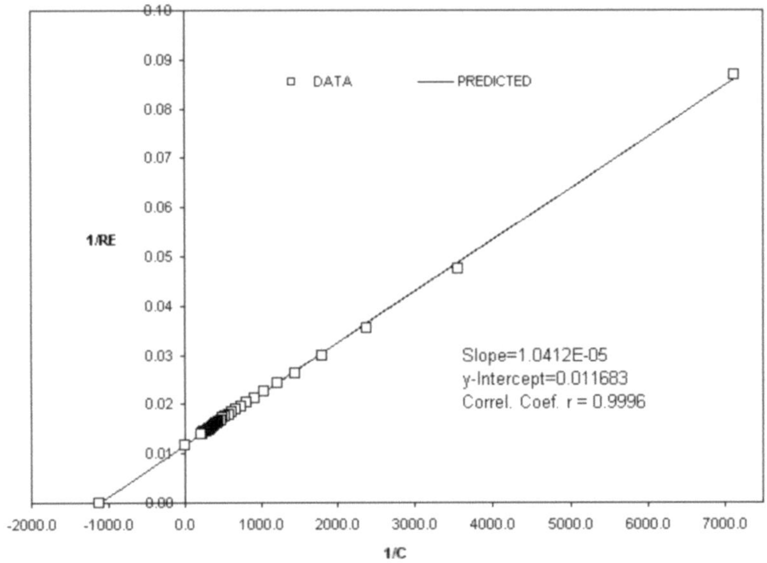

Fig. 11. Lineweaver-Burk plot for PANI-valinomycin modified electrode

The perfect fit of the titration data to the Michaelis-Menten model as shown on Figure 11, is a clear evidence of facilitated diffusion mechanism on the assembled biomimetic membrane on GC electrode.

9. Analogy to enzyme kinetics Michaelis-Menten equation

Ion carrier proteins exhibit Michaelis-Menten kinetics (Nelson 2000), and their affinity for specific ions can be measured in terms of two parameter, the Michaelis constant K_m, which measures the affinity for the ion carrier as an analogy to enzyme-substrate interaction, and the V_{max} which measures the maximal velocity of the reaction at saturating ion concentration as represented by the rectangular hyperbolic curve.

In enzymatic reactions the slowest step is the conversion of [ES] to [E] + [P] , and in such cases the value of k_{cat} is much less than k_2, thus $K_m \cong k_2/k_1=K_d$ the dissociation constant for binding of S to E, thus the value of K_m describes the affinity of an enzyme for its substrate, and hence the stability of the enzyme-substrate complex. The smaller the value of K_m the more avidly the enzyme can bind the substrate from a dilute solution and the smaller the concentration of substrate needed to reach half-maximal velocity.

A Lineweaver-Burk plot is a graphical representation of enzyme kinetics based on the plotting of double reciprocals that facilitates the interpretation of the Michaelis-Menten plot.

The Lineweaver-Burk double reciprocal plot is notorious for distorting data, and it is unreliable for the calculation of enzyme kinetics parameters (Rose 2006), however it is very useful as a graphical representation of the different types of enzyme inhibitions, e.g. it can used to distinguish between competitive, noncompetitive and uncompetitive inhibitors. In competitive inhibition one can note that inhibitors have the same y-intercept ($1/V_{max}$) as uninhibited enzyme but different slopes (K_m/V_{max}) and x-intercepts ($-1/K_m$) are observed for the data sets.

For noncompetitive inhibition the curves have the same x-intercept but different slopes and y-intercepts. Finally in uncompetitive inhibition the curves have different intercepts on the y and x axes, however the slopes are the same.

Diffusion is a mean of passive transport; it results from the thermal, random movement of ions and molecules. There are three main types of diffusions; simple, channel, and facilitated diffusion. Simple diffusion does not involve a protein and occurs when a small, non-polar molecule passes through a lipid bilayer without being rejected. Channel diffusion involves channel proteins where material moves through an open aqueous pore. Facilitated diffusion is dependent on single ion moving along ionophores that operate on a bind, flip, and release mechanism. This type of diffusion mechanism allows hydrophilic molecules to move into the hydrophobic region of the membrane without being rejected.

The valinomycin modified lipid bilayer assembled on PANI works by selectively complexing the K^+ ions from their Cl^- counterions in the test sample solution, thereby causing a charge separation and a corresponding change in electrical conductivity in the membrane. This causes the generation of an EMF, which could be defined as the force with which positive and negative charges could be separated and it represents the potential value in the Nernst equation.

The electroconductive PANI film which is supporting the lipid membrane acts as a good ion-to-electron transducer. Because valinomycin is selective on its affinity for K^+ ions any measurable change in potential is due solely to the presence of this ion. In summary the ionophore valinomycin has an effect on the electrical properties of phospholipid bilayer membranes, such that it effects the solubilization of K^+ ions within it, thereby providing a carrier mechanism by which this ion can cross the insulating hydrophobic or hydrocarbon interior of the bilayer. When accumulated in the lipid membrane assembled on PANI electrode they cause a voltage differential which can be determined using an external reference electrode. Because only K^+ binds to the valinomycin in the membrane, the conductive path only appears for K^+. Therefore, the potential developed is attributable solely to the K^+ concentration. The electrical neutrality of the lipid membrane is maintained by a reverse flow of H^+.

10. Potentiometric titration using gramicidin PANI modified electrode

A thiolated PANI GC electrode was coated with a suspension of the pentadecapeptide gramicidin (mw 1882.3 g mol^{-1}) in PC LUVs, and allowed to equilibrate at the physiological temperature of 37°C. The procedure is shown below:

Pipetted 2 mL of egg L-α-Phosphatidylcholine (with a concentration of 20 mg/mL in chloroform) into a glass scintillation vials and evaporated under a stream of nitrogen, then kept under vacuum for 1-hour. Added 2.8 mg of gramicidin, followed by 15 mL of 0.5M H_2SO_4 and vortexed for 5 minutes to make a suspension. Transferred an aliquot onto a 10x75mm glass centrifuge tube and carefully suspended one thiolated PANI GC electrode inside the tube with the aid of Parafilm® such that it did not touch the bottom of the tube. Placed in a controlled temperature bath for 60 minutes at 37°C, removed gently from the tube, and slowly washed the PTFE electrode body with Milli-Q water, never touching the bottom of the electrode. Special care was taken to protect the assembled membrane by suspending the electrode on the electrochemical cell lid and equilibrating in 10 mL of Milli-Q water until a steady potential reading was reached.

The titration experiment was conducted in a glass electrochemical cell containing 10 mL of Milli-Q water and outfitted with a three holed PTFE lid to support the gramicidin modified electrode, and the Ag/AgCl reference electrode, with the third hole left open to add the titrant solution. The two electrodes were connected with a BNC cable to a pHmeter operating in potential mode. The electrodes were allowed to equilibrate until reaching a steady absolute potential reading, and then switched to relative potential mode thus displaying a zero value before starting the addition of titrant.. The titration results are shown on Appendix D and are depicted graphically in Figure 12 and 13.

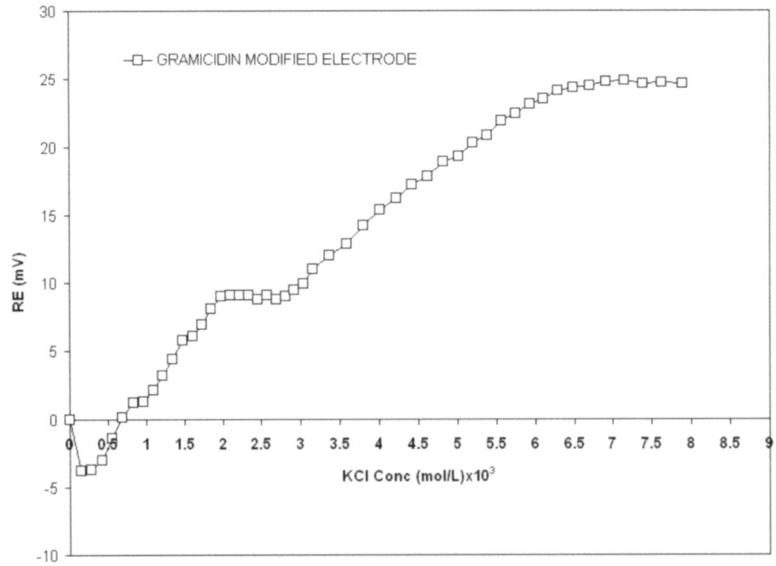

Fig. 12. Potentiometric titration curve for PANI-gramicidin modified electrode

The electrode showed a change in response at a concentration of about 0.2 mg/mL of KCl as shown on Figure 12. The trend line shown on Figure 13 includes all the data point, however a close examination of this attempted fit to a Nernstian plot, revealed what appeared to be two linear sections of different slopes separated by a horizontal segment, an a clear evidence of an apparent mixed bimodal titration mechanism as shown on Figure 14. The first segment has a smaller slope than the second segment, indicating than the binding of K^+ ions was initially slow, and then changed after reaching a concentration of approximately 0.2 mg/mL, increasing by a factor of approximately 4. The gramicidin molecule has physical dimensions[42] of 30Å-long by 15Å-wide, while the valinomycin molecule dimensions are 5.6Å-long by 15.6Å-wide. For this reason the gramicidin molecule spans about 5 times deeper into the lipid bilayer compared to valinomycin.

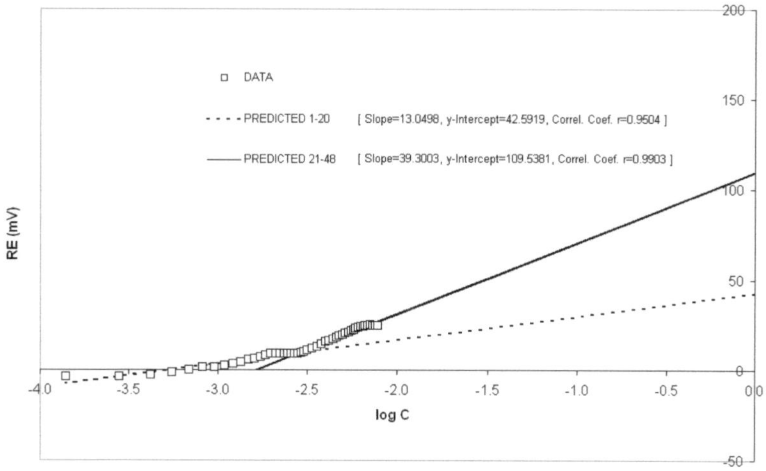

Fig. 13. Nernstian plot for potentiometric titration with PANI-gramicidin modified electrode

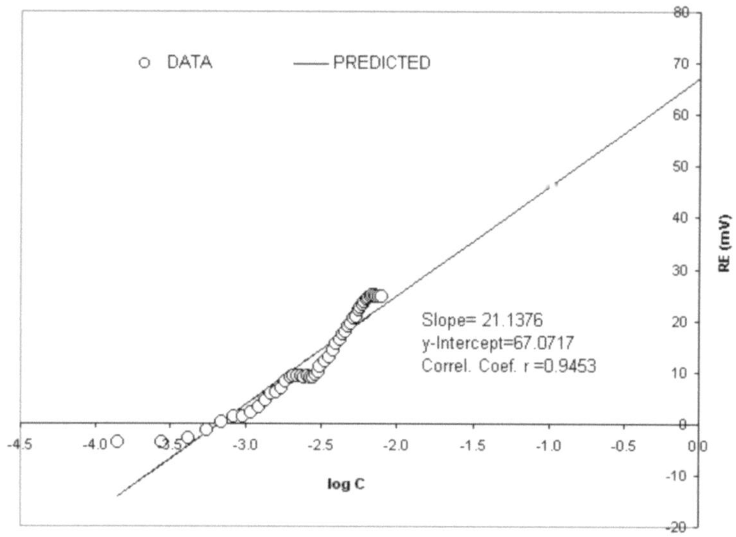

Fig. 14. PANI-gramicidin modified electrode mixed bimodal titration mechanism

The bimodal titration mechanism may be due to the fact that assembled simulated were not tethered. Tethered bilayer lipid membranes (tBLMs) offer a quasi-natural environment where membrane proteins can be embedded in and investigated. A tBLM consists of a lipid bilayer that is coupled covalently via a spacer group to a solid support on an electrode, offering homogeneity and fluidity that is important toward the biomimesis of biological membranes, and also helps accommodate larger proteins.
The suspension of gramicidin and valinomycin in PC allowed the assembly of the ionophores on the derivatized phospholipid layer on PANI GC electrodes. The ion carriers settled in the assembled phospholipid layers creating hydrophilic pores which can open and allow inorganic ions, to pass through. In general, ion carriers are quite specific for the type of solute they will transport and it occurs very fast by a mechanism called facilitated diffusion. Facilitated diffusion is a process of diffusion, and a form of passive transport that facilitates spontaneous passage of molecules or ions across a biological membrane passing through specific transmembrane transport proteins.

11. Conclusions

A successful assembly of biomimetic membranes on phospholipid derivatized PANI-GC electrode was achieved by using the ion carrier proteins gramicidin and valinomycin. The valinomycin assembled electrode showed a titration curve consistent with facilitated diffusion mechanism as this carrier has been shown to have a great affinity for the K^+ ion. Evidence of such affinity was provided by a perfect fit of the titration data to a Michaelis-Menten enzyme-substrate model. The electrode showed a great detection slope of approximately 42 mV/decade.
The gramicidin modified electrode showed a bimodal titration mechanism which was probably caused by its larger size as compared to valinomycin. No tethers were used to attach the simulated membrane to the electrode. The slope of the first segment of the titration curve was approximately 13 mV/decade, and after reaching a KCl concentration of approximately 0.2 mg/mL, the simulated membrane appeared to swell to the point of at which its fluidity improved, then the sensitivity of the electrode increased to a slope value of 39.3 mV/decade. It appeared that valinomycin is a better ion carrier to bind K^+ ions for this type of electrode, as it showed enough fluidity, and a good sensitivity without the need of tethers

12. References

K. Ataka, F. Giess, W. Knoll, R. Nauman S. Haber-Pohlmeier, B. Richter and J. Heberle, J Am Chem Soc 126 16199-16206 (2004)
E. Bakker, E. Pretsch, Trends in Analytical Chemistry, 27(7), 1-7, (2008)
T. M. Baryerl, M. Bloom, Biophys. J., 58, 357-362, (1990)
C. D. Batich, H. A. Laitinen, and H. C. Zhou, J. Electrochem. Soc. 137, 883 (1990)
P.M. Beadle, Y.F. Nicolau, E. Banka, P. Rannou, D. Djurado, Synth, Met. 95, 25 (1998)
J. Bobacka,, A. Ivaska, A. Lewenstam, Analytica Chimica Acta 385 195-202 (1999)
C. L. Brosseau, J. Leitch, X. Bin, M. Chen, S.G. Roscoe, J. Lipkowski, Langmuir 24, 13058-13067 (2008)
A. Cadogan, A. Gao, A. Lewenstam, A. Ivaska, D. Diamond, Anal. Chem., 64, 2496, (1992).
G. Ćirić-Marjanović, E. N. Konyushenko, M. Trchová, J. Stejskal. Synth. Met. 158, 200 (2008)

J. Desilvestra, W. Scheifele, and O. Haas, J. Electrochem. Soc. 139, 2727 (1992).
F.B. Diniz, K.C.S De Freitas, W.M. De Azevedo, Electrochemistry Communications, 1, 271-273, (1999)
E. M. Genies, A. Boyle, M. Lapkowski, and C. Tsintavis, Synth Met. 36, 139 (1990).
J.T. Groves, M.L. Dustin, J. Immunol. Meth., 278, 19-32 (2003)
C-C Han, S-P Hong, K-F Yang, M-Y Bai, C-H Lu, and C-S Huang, Macromolecules, 34, 587 (2001)
C-C Han, R-C Jeng, Chem. Commun, 553 (1997)
Z. Jin, Y. Su and Y. Duan, Synth. Met. 122 (2), 237-242 (2001)
A. Kabumoto, K. Shinozaki, K. Watanabe, and N. Nishikawa, Synth. Met. 26, 349 (1988)
E. Kalb, S. Frey, and L.K. Tamm, Biochim. Biophys. Acta, 1103, 307-316, (1992)
E. Kalb, J. Engel, and L.K. Tamm, Biochemistry, 29, 1607-1613, (1990)
H. Karami. M. F. Mousaui, Talanta, 63, 743-749, (2004)
K. Kjaer, J. Als-Nielsen, C.A. Helm, L.A. Laxhuber, and H. Mohwald, Phys Rev. Letter, 58, 2224-2227, (1987)
P. T. Kissinger, W. R. Heineman, Laboratory Techniques in Electochemical Analysis, 2nd Edition, Marcel Dekker, Inc., New York, (1996), Chapter 2.
J. Lipkowski, Physical Chemistry Chemical Physics 12(42) 13853–14368 (2010)
L. V. Lukachova, E. A. Shkerin, E. A. Puganova, E. E. Karyakina, and A. A. Karyakin, J. Electroanal. Chem. 544, 59 (2003).
J Majewski, J Y Wong, C K Park, M Seitz, J N Israelachvili, and G S Smith, Biophys J. 75(5), 2363-2367, (1998)
R. Mazeikiene, A. Malinauskas, Synth. Met., 89, 77-79, (1997)
A. Merz, A. J. Bard, J. Am. Chem. Soc., 100, 3222, (1978)
L. L. Miller, M. R. Van de Mark, J. Am. Chem. Soc., 100, 3223, (1978)
H. Naarman and N. Theophilou, Synth. Met. 22, 1, (1987)
R Naumann, D. Walz, S. M. Schiller, W. Knoll Journal of Electroanalytical Chemistry, 550-551, 241-252 (2003)
D.L. Nelson and M.M. Cox, Lehninger's Principles of Biochemistry, 3rd Edition, Worth, New York, (2000)
P.C. Pandey, A.P. Mishra, Sensors and Actuators B, 99, 230-235, (2004)
L. Rose, A.T.A. Jenkins, Bioelectrochem., 71, 114-120, (2006)
J. F. Rusling, Acc. Chem. Res., 24, 75-81, (1991)
E. Sackmann, Science, 271, 43-49, (1996)
C. Slim, N. Ktari, D. Cakara, F. Kanoufi, and C. Combellas, J. Electroanal. Chem., 612, 53-62, (2008)
J. Stejskal, R.G. Gilbert, Pure Appl. Chem. 74(5), 857-867 (2002)
T. Sulimenko, J. Stejskal, I. Krivka, J. Prokes, Eur. Polym. J. 37, 219 (2001)
M. Tanaka, E. Sackmann, Nature, 437, 656-663, (2005)
S.A. Tatulian, P. Hinterdorfer, G. Barber, and L.K. Tamm, EMBO J. 14, 5514-5523, (1995)
S. Terretaz, T. Stora, C. Duschl, and H. Vogel, Langmuir, 9, 1361-1368, (1993)
A.P.F. Turner, I. Karube, G.S. Wilson, Biosensors, Fundamentals and Applications, Oxford University Press, USA, (1990)
A. Varela, and R. M. Torresi, J. Electrochem. Soc. 112, 2800 (1990)
M. Vazquez, Potentiometric Ion Sensors Based on Conducting Polymers, Doctoral Thesis, Åbo Akademi University, Åbo, Finland, (2005)

J.Y. Wong, J. Majewski, M. Seitz, C.K. Park , J.N. Israelachvili, G.S. Smith, 1: Biophys J., 77(3), 1445-1457, (1999).
D. Zhang, Polymer Testing, 26, 9-13 (2001)
X. Zhang, B. Oborevc, J. Wang, Analytical Chimica Acta, 452, 1-10, (2002).

5

Biomimetic Sensing – Actuators Based on Conducting Polymers

Joaquín Arias-Pardilla[1], Toribio F. Otero[1], José G. Martínez[1] and Yahya A. Ismail[2]
[1]*Universidad Politécnica de Cartagena*
[2]*University of Nizwa*
[1]*Spain*
[2]*Sultanate of Oman*

1. Introduction

Thirty years after the discovery of conducting polymers (CP) and their electrochemical behavior (Chiang et al. 1977; Shirakawa et al. 1977), CP have been proven to be useful for applications in a variety of areas such as batteries and supercapacitors, photovoltaics, memory devices, light emitting diodes, artificial muscles (actuators), biomedical devices, biosensors, corrosion protection etc., thanks to their unique physical and electrochemical properties. Their physical characteristics are similar to those of commodity polymers; but unlike commodity polymers, CP can be subjected to oxidation and reduction just as in the case of metals and inorganic semiconductors. These reactions yield a complex material combination: polymer-ions-solvent.

In the field of actuators, conductive polymers have been used with great success. During its oxidation/reduction the volume of the CP film changes. The volume variation can be used to perform a linear movement or an angular movement. The first CP actuators date back to 1992 (Baughman & Shacklette 1991; Otero et al. 1992a; Pei & Inganäs 1992b). Since then, much progress has been achieved in materials generation and design and characterization of devices. Different configurations have been designed and studied, and different models were proposed to understand the properties and to improve the performance of those devices. In addition, due to the electrochemical nature of the actuation, any variable acting on the reaction will be detected and quantified by the actuator response; it means that working under constant current, the potential evolution will detect the change. This is a unique device: artificial muscles based on conducting polymers are also sensors, while working, of the environmental variables (Otero 2009). They are sensors of temperature, electrolyte concentration and the current flowing through them. Since they are capable of detecting an obstacle along its route and moving it to performing a work, they possess tactile properties (Otero & Cortes 2003).

These unique features, coupled with the easy construction of actuators in many different sizes, from macro to microscopic devices, constitute a technological challenge. Some multinational companies, like Bayer, became involved in this new technology after acquisition of startup companies. The European Union is supporting the European Scientific

Network for Artificial Muscles (ESNAM). Development of new actuators is an ongoing active topic of research for the ESA and NASA, the European and American space agencies.

2. Electrochemical reactions in conducting polymers

Much of the importance of these materials emanates from the fact that they can be oxidized and/or reduced from its neutral state, with the entry/exit of ions and solvent (reactions 1 and 2). The chemical reaction starting from the neutral polymer promotes an increase of the electrical conductivity and a change of material composition. With the change in composition, many other properties also change (they are the reactive properties or electrochemical properties), as can be seen in Table 1 (Otero 1999; Otero 2000).

Property	Action	Inspired organ
Electrochemomechanical	Change of volume	Muscles
Electrochromic	Change of color	Mimetic skins
Charge storage	Current generation	Electric organs
Chemical or pharmacological storage	Chemical modulation or chemical dosage	Glands
Electron/neurotransmitter	Channel V action	Nervous interface
Electroporosity	Transverse ionic flow	Membrane
Electron/ion transduction	ΔV (Chemical/Physical properties)	Bio-sensors

Table 1. Biomimetic properties of conducting polymers driven by the electrochemical reaction, the mimicked biological functions and the related organs.

The reactive material is composed of reactive macromolecules, ions and solvent: this composition mimics dense reactive gels in the living cells, becoming reactive material models for biological reactions. The reverse variation of the magnitudes of biomimetic properties driven by the reverse electrochemical reactions envisages the development of new biomimetic (electrochemical) devices and products like artificial muscles, electrochromic windows, mirrors or screens (Dyer & Reynolds 2007; Mortimer et al. 2006; Rosseinsky & Mortimer 2001), polymeric batteries (Irvin et al. 2007; Novak et al. 1997), and/or supercapacitors (Peng et al. 2008; Snook et al. 2011), smart drug delivery devices (Abidian et al. 2006; Kontturi et al. 1998; Pernaut & Reynolds 2000; Zinger & Miller 1984), nervous interfaces (Cui et al. 2001; Ludwig et al. 2006; Schmidt et al. 1997), smart membranes (Ariza & Otero 2007; Burgmayer & Murray 1982; Pellegrino 2003) or smart surfaces (Isaksson et al. 2006; Isaksson et al. 2004; Teh et al. 2009). A breaking technological border has appeared giving rise to reactive biomimicking materials and reactive devices based on properties whose magnitude changes along with the progress of material reaction or along the device actuation.

The reactions that we describe below are simplified expressions and initial approach to real processes. Any film of a CP acts as a polymeric membrane. As for any other membrane inside a liquid electrolyte, a physical equilibrium is stated between the film and the electrolyte: solvent, anions and cations getting a distribution between the polymer and the electrolyte. Then, an interchange of ions and solvent starts with the beginning of the electrochemical reactions (Hillman et al. 1989; Inzelt 2008). For most of the polymers one of the two ionic exchanges prevails carrying over 90% of the balancing charge (Orata & Buttry 1987; Torresi & Maranhao 1999).

2.1 p-doping

P-doping, generally known as oxidation, involves the extraction of electrons from the polymer chains, generating positive charges, or holes, along the polymer chains, which are compensated by ions coming from the electrolytic medium in contact with the polymer. This type of doping is the most common in CP (Huang et al. 1986; Tsai et al. 1988). The transformation from a neutral material to an oxidized material can occur through two different ways:

- **Prevailing anion exchange**: During oxidation, stimulated conformational movements on the polymeric chains generate free volume and anions penetrate from the solution to balance the positive charges along the chains. The high concentration of charges inside the oxidized material induces the entrance of solvent from the solution required to maintain the osmotic pressure. The volume of the material increases during oxidation. Reverse processes and a volume decrease is observed during neutralization, as observed in reaction (1):

$$\left(Pol^{0}\right)_{s} + n\left(A^{-}\right)_{aq} + m\left(Solv\right) \rightleftarrows \left[\left(Pol^{n+}\right)_{s}\left(A^{-}\right)_{n}\left(Solv\right)_{m}\right]_{gel} + n\left(e^{-}\right)_{metal} \quad (1)$$

Neutral chains Oxidized chains

where the different sub indexes mean: s, solid and aq, solution, Pol^0 represents the reactive centres along the chains where a charge can be stored and A^- represents the anions. Volume changes and the transition, promoted by the reaction, from a packed solid state to a gel are illustrated in Fig. 1. The polymer film acts as a three dimensional gel electrode at the molecular level. Inside the swelling film every chain constitutes a molecular one-dimensional electrode. All the chains have the same chemical potential (μ) at any electrical potential.

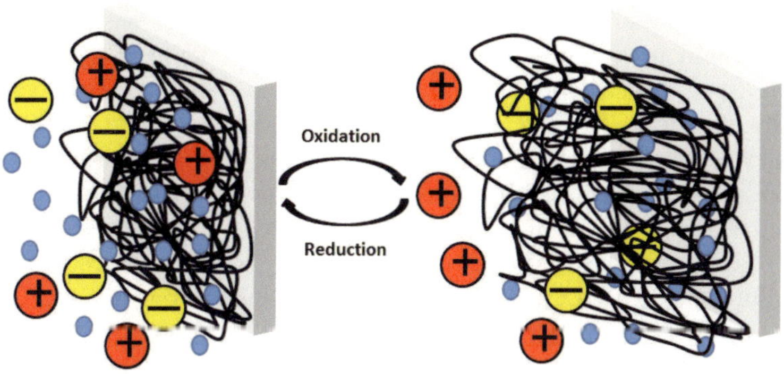

Fig. 1. Complex polymer structure, showing volume changes (swelling/shrinking) produced during p-doping of polypyrrole film. Anions go inside the film during oxidation and are released during neutralization. Blue circles represent solvent molecules.

- **Prevailing cation exchange**: If the polymer film is generated in presence of a big enough macro anion, this anion remains trapped inside the polymer matrix whatever may be the oxidation state. In this case when the polymer is oxidized, balancing cations are expelled as shown in Eq. 2. The material shrinks during oxidation and swells during reduction.

$$\left[\left(Pol^{0}\right)\left(MA^{-}\right)_{n}\left(C^{+}\right)_{n}\right]_{s} \rightleftarrows \left[\left(Pol^{n+}\right)\left(MA^{-}\right)_{n}\right]_{s} + n\left(C^{+}\right)_{aq} + n\left(e^{-}\right)_{metal} \quad (2)$$

Neutral chains Oxidized chains

where MA^- represents any macroscopic anion (organic, polymeric or inorganic) trapped inside the CP during polymerization and C^+ represents a cation. Whether macroanions during reaction (2) remain trapped inside the film by its relative dimensions with the interchain distances in the network, or by strong intermolecular Van der Waals interactions between the macroion and the polymer is an unsolved point. The role of the solvent molecules in this case is not clear due to the presence of ionic species in the material, having strong interactions with the solvent dipole, whatever their oxidation state would be.

2.2 n-doping

Some CP such as PEDOT (Ahonen et al. 2000; Skompska et al. 2005), polythiophene (Arbizzani et al. 1995) or polyfluorenes (Ranger & Leclerc 1998) have an electronic affinity high enough to allow transitions from the neutral state to a reduced state, enabling them to store negative charges (by electron injection) on the chains at high cathodic potentials. In this case very stable solvents and salts are required, as electrolytes, to perform this reaction (3). The material swells during reduction and shrinks during neutralization.

$$\left(Pol^{0}\right)_{s} + n\left(C^{+}\right)_{aq} + n\left(e^{-}\right)_{metal} \rightleftarrows \left[\left(Pol^{n-}\right)_{s}\left(C^{+}\right)_{n}\left(Solv\right)_{m}\right]_{gel} \quad (3)$$

Neutral chains Reduced chains

3. Macroscopic dimensional changes

Though the polymer chains could be an example of molecular motors (Balzani et al. 2005; Davis 1999) due to their change in length (conformational changes) during oxidation or reduction, it is not yet possible to exploit this phenomenon because we are not able to connect each end of the chain to a nanoscopic contact taking advantage of that anisotropic linear expansion. Only three dimensional films of CP, giving three dimensional changes of volume are available. By chemical or electrochemical polymerization we produce films of branched, partially cross-linked and partially degraded materials. This entanglement of molecular motors swells or shrinks under electrochemical stimulation of the conformational movements of the electroactive part of the chains. The result is a quite isotropic three dimensional change of volume to produce mechanical properties. In order to generate macroscopic machines from the isotropic volume changes in the film we must induce anisotropy.

4. Classification of actuators: Type of movement

The actuators made of CP can be classified in several ways, one of them is based on the device design to produce different types of macroscopic movements.

4.1 Bending structures

4.1.1 Bilayer

This was the first type of CP actuator that was developed in 1992 by the construction of bilayers constituted by a tape adhered to a film of a CP (Fig. 2) (with prevalent anion exchange, or cation exchange) electrogenerated on a metallic electrode (Otero et al. 1992b; Otero et al. 1992a; Otero et al. 1992c; Pei & Inganäs 1992b). The mechanical stress gradient generated across the bilayer interface by swelling /shrinking processes induced by the

electrochemical reactions develops a macroscopic movement of the bilayer free end by the device bending. Films of CP having a prevailing exchange of anions swells by oxidation, pushes the bilayer free end and stays at the convex side of the bended device. CP having cation prevailing exchange suffer shrinking processes during oxidation trailing the device and staying at the concave side of the bended device. Different materials have been used to prepare bilayer devices as CP/metal (Jager et al. 2000a; Jager et al. 2000b; Smela et al. 1993), CP/solid state electrolyte(Baughman 1996), CP/CP (Han & Shi 2004a; Takashima et al. 2003a), CP/plastic(Higgins et al. 2003), CP/paper (Deshpande et al. 2005b) or CP/thin film of any flexible material which is metal coated (i.e. by sputtering) (Deshpande et al. 2005a).
In these devices, a metallic counter electrode is required to allow the current flow. A major fraction of the consumed electrical energy is wasted to produce electrochemical reactions on this counter electrode (such as solvent dissociation, which requires a high overpotential). Moreover, those reactions result in pH variations and new chemicals, which migrate towards the muscle promoting the progressive deterioration of the actuating film.

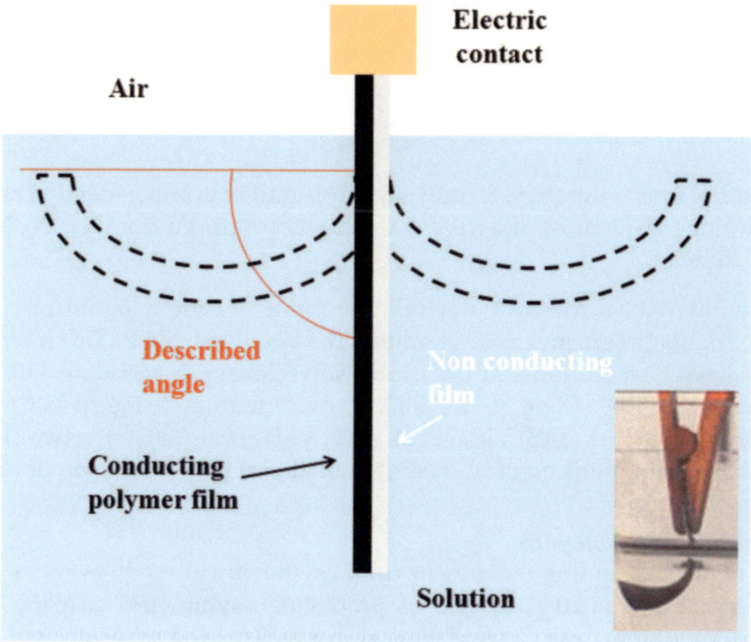

Fig. 2. Bilayer device scheme in solution, formed by a conducting polymer film and a non-conductive film. A real device is shown in inset.

4.1.2 Three-layer structure

This configuration evolves from the bilayer, trying to avoid the metallic counter electrode (Fig. 3) getting higher efficiencies by using the same current twice to produce opposite volume changes. This is produced by using a two sided polymeric tape and two films of the same, or different, CP which are allowed to stick on both sides of the tape (Otero et al. 1992c). The triple layer is immersed in an electrolyte allowing the current flow. One of the CP films acts as the anode, swelling (for preferential anionic interchange during the reaction) and pushing the device while the second CP film acts as the cathode, shrinking and trailing the device (John et al. 2008; Yao et al. 2008).

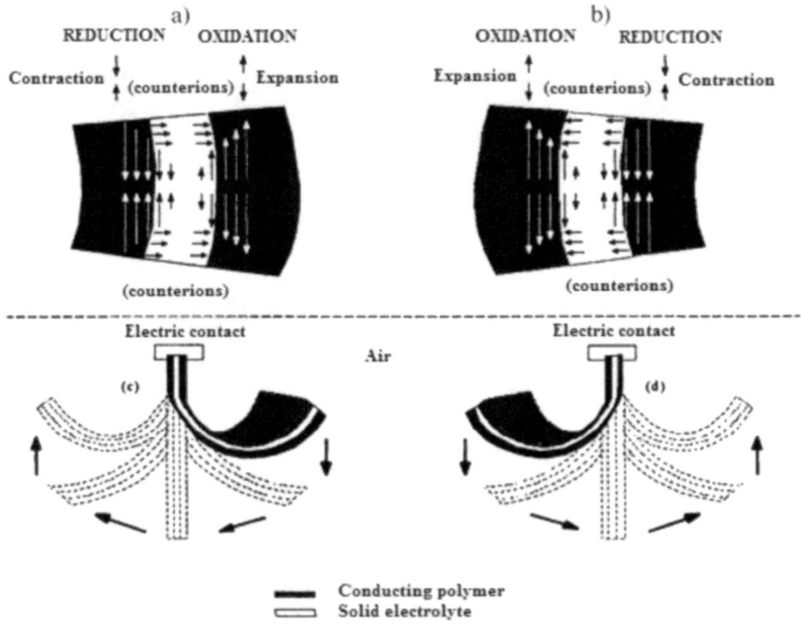

Fig. 3. A scheme of ionic interchange, induced stress gradients and generated angular movements during current flow of a trilayer device (Reproduced from (Otero 2000), with permission of Elsevier).

But unlike the bilayers, three-layer devices can move outside a liquid electrolyte media using an ionic conducting membrane separating the two films of CP. This membrane can be obtained by solvent evaporation and UV irradiation (Blonsky & Meridian 1997; Heuer et al. 2002; Sansinena et al. 1997; Song et al. 2002), or by formation of interpenetrated networks (Cho et al. 2007; Plesse et al. 2005; Vidal et al. 2009; Vidal et al. 2003): the two films of CP are generated by chemical polymerization on the external part of the film membrane.

4.1.3 Asymmetrical monolayers

It is possible to obtain bending movement from asymmetrical monolayers of the same CP, having an internal asymmetry capable of producing asymmetric swelling or shrinking across the film under the same electrochemical process (doping or de-doping) (Okamoto et al. 2000; Onoda et al. 1999a; Onoda & Tada 2004; Onoda et al. 1999b; Shakuda et al. 1993; Takashima et al. 2003b; Takashima et al. 1997; Wang et al. 2002): one half of the film has a prevalent anionic exchange, while the second half experiences a prevalent cationic exchange. This can be achieved in two separate stages of electrogeneration using different dopants for the same polymer. A metallic counter electrode is required.

Other possibilities are being explored to produce asymmetric monolayers by physical means, for example, by growing the CP on adsorbed and porous materials (Li et al. 2004), or by electrochemical means generating a film of CP with a counterion concentration gradient (Okuzaki & Hattori 2003; Sansiñena et al. 2003; Shakuda et al. 1993; Wang et al. 2002), conductivity (Nakano & Okamoto 2001; Onoda et al. 2005) or morphology gradients (Han & Shi 2006; Okamoto et al. 2001; Onoda & Tada 2004) by crosslinking network, or by generating a bilayer of the conducting polymer with a macroanion (shrinks by oxidation) and then of the same conducting polymer with a small anion (swells by oxidation) (Han & Shi 2006; Takashima et al. 2003a), or even placing a metal sheet between both films (Han & Shi 2004b). As in the case of bilayer devices, asymmetric films also need an electrolyte and a metallic counter electrode in order to allow the flow of the current and to produce the bending movement.

4.2 Structures giving lineal movements

4.2.1 Films and fibers

A freestanding CP film is the simplest lineal actuator possible (DellaSanta et al. 1997a). Its actuation principle is based on longitudinal expansion and contraction of the polymer during the insertion and de-insertion of ions although expansion and contraction occur in all three dimensions, as indicated earlier. Different strategies were used to improve the performance of these actuators, like depositing platinum on CP to increase conductivity (Hutchison et al. 2000); although the resulting films are very brittle and difficult to handle. To overcome these problems multilayered actuators were proposed, in which several thin laminations of CP films and an electrolyte (ionic liquid-soaked paper) are used to develop a compact and scalable linear CP actuator with a high work output (Ikushima et al. 2009). Also folded films with Origami shapes provide good linear movements (Okuzaki 2008).

Different methods are being used to obtain fibers of CP. One of them is extrusion through a spinneret of a polymer soluble organic solution (normally dimethylpropylene urea (DMPU) or N-Methyl-2-pyrrolidone (NMP)), followed by a couple of stages, first being an immersion in water followed by immersion in doping bath (Mazzoldi et al. 1998). Another method is chemical polymerization over a fiber-shaped substrate (Lu et al. 2002). This substrate can be a hollow fiber solid polymer electrolyte (Plesse et al. 2010) making it possible to obtain two concentric CP films separated by the electrolytic medium, allowing its movement in air.

Bundles of films or fibers were investigated to produce vertical displacements of weights (Lu et al. 2002). Different models have been proposed to describe mechanical behaviour (DellaSanta et al. 1997b; DellaSanta et al. 1997c; Qi et al. 2004) or electro-chemomechanical deformations (Kaneto et al. 2000; Pandey et al. 2003; Zama et al. 2004) of these devices. When individual fibers or bundles are used as working electrodes, a metal counter electrode is required to allow the current flow generating similar effects described above for bending asymmetric films.

4.2.2 Tubes and films with metal support

A different approach consists of electropolymerization of a CP on springs and helical metallic wires until the generation of a tube (Ding et al. 2003; Hara et al. 2003; Hara et al. 2004b; Spinks et al. 2003b), or on zigzag metal wires (Hara et al. 2004a) to generate films. The supporting metal wire should guarantee a uniform potential and current distribution, greater strain and length variation rates under electrochemical reactions.

4.2.3 Combination of bending structures

Another way to obtain a linear displacement is the combination of different bending structures as bilayers (Otero & Broschart 2006) or trilayers (Otero et al. 2007; Otero et al. 2002) capable of achieving a longitudinal displacement of up to 60% of their original longitude. In the case of longitudinal displacements where small bending angle lower than a 20% is required, the films suffer low mechanical fatigue.

5. Conducting polymer actuator applications

While the investigation of these devices is still mainly in academic laboratories, some companies dedicated to the development of actuators based on conjugated polymers have emerged in recent years. MicroMuscle AB based in Sweden which was later acquired by Creganna Tactx Medical, Artificial Muscles from California acquired by Bayer MaterialScience and EAMEX from Japan are actively working on actuators for biomedical and electronics applications. Also Santa Fe Science and Technology, USA, has produced continuous spun polyaniline fibres and demonstrated their use as linear actuators. Many different applications can be found in literature. The following is a summary of a few of them, both macroscopic and microscopic.

5.1 Propulsion or locomotion systems

Biomimetic systems based on tri-layer polymer combination of actuators are proposed for swimming devices (Alici et al. 2007), where undulatory movements were employed to create enough thrust for propulsion. Bending actuators like cilia can be used in biological systems to produce movement in mini-robotic systems capable of performing applications like pipe inspection, search inspection and data gathering in confined spaces (Alici & Gunderson 2009).

5.2 Braille displays

CP is one of the technological materials explored to develop low-cost, efficient, and refreshable displays for Braille text. Producing a full-page Braille display is difficult because it requires packing of many small actuated dots into a small, closely-spaced arrangement without interferences. Actual displays are close to performing at the required specs; challenges still limit their commercial viability such as short cycling life (Ding et al. 2003; Spinks et al. 2003a).

5.3 Steerable microcatheters

Two CP films around a passive catheter and a fast ionic conductor solid polymer electrolyte are used to control the movement of the catheter tip, transforming passive catheter to an active and steerable one. Such devices are typically expected to produce a bending angle of at least 20° (Santa et al. 1996; Shoa et al. 2008). CP are also being used in Optical Coherence Tomography fast scanning catheters to enable high-resolution 3-D imaging (Lee et al. 2009).

5.4 Integrated cell sensors

Cell-based sensors are being developed to harness the specificity and sensitivity of biological systems for sensing applications, from smell detection to pathogen classification (Jager et al. 2002; Smela et al. 2007). These integrated systems consist of CMOS chips containing sensors and circuitry by means of which microstructures have been fabricated to transport, contain, and nurture the cells. The structures for confining the cells are micro-vials that can be opened and closed using polypyrrole (pPy) bilayer actuators.

5.5 Micro pump

Volume change of CP can be used to develop a micro pump system capable of transporting fluids at a microflow rate with high precision. Micropumps are essential components of microfluidic systems and biosensing systems. The energy consumption rate of these polymeric micro pumps is markedly lower than those of conventional micropumps (Naka et al.; Ramírez-García & Diamond 2007; Wu et al. 2005).

5.6 Micro actuators

The electrochemical synthesis of CP films and their electrochemical actuation are suitable for the construction of elegant and imaginative microdevices (He et al. 2007; Jager et al. 2001; Pede et al. 1998; Roemer et al. 2002) and microtools constituted by bilayers (Jager et al. 2000a; Jager et al. 2001; Jager et al. 2000b; Jager et al. 1999; Smela 1999; Smela & Gadegaard 1999; Smela et al. 1995; Smela et al. 1993) or trilayers (Kiefer et al. 2008) using microelectronic technologies.

6. Models for the characterization of the movement

Throughout the development of these new devices, researchers have proposed different models to characterize their movement, using different approaches, from physical and mathematical methods to others based on their physicochemical properties.

6.1 Faradic control of the movement

All the above described artificial muscle structures can be considered as electrochemical devices working under faradic conditions. The volume changes between reduced and oxidized polymer films are under control of the used charge, either injected or extracted. This electrical charge controls the amount of counterions interchanged with the solution, and the volume change is related to the stress gradient variations at the polymer/non-conducting tape interface. Experimentally, different electrical charges (Q) applied to the device produce proportional movements (α) (Eq. 4). The constant, k, is a function of the nature of ions exchanged in the process.

$$\alpha\ (\mathrm{rad}) = k\ (\mathrm{rad \cdot s^{-1} \cdot A^{-1}}) * Q\ (\mathrm{A \cdot s}) \quad (4)$$

From this equation, a linear relationship is obtained between applied current (I) and the rate of the angular movement, ω, (Eq. 5).

$$\omega\ (\mathrm{rad \cdot s^{-1}}) = k\ (\mathrm{rad \cdot s^{-1} \cdot A^{-1}}) * I\ (\mathrm{A}) \quad (5)$$

Eq. 5 confirms that all these electrochemomechanical actuators are electrical machines, whose movement is under perfect control of the driving current (magnitude and sense of flow). Whatever the initial position of the device free end is, the charge required to produce an angular movement of one radian must be constant, and this amount is a characteristic magnitude of the tested system. Therefore, the movement can be accelerated using a higher current and retarded by current decrease. The movement stops when the current flow is halted, and the movement sense is reversed by changing the current sense.

The electrochemical nature of these devices allows to anticipate the response of different devices, having different surface areas, different shapes or including different polymer mass per device. The angular rate produced by interchange of the same number of moles of ions (cations or anions) per unit time and per unit mass of CP during actuation is the same and independent of the shape, surface area or amount of conducting polymer in the device. That means the same change of the specific composition (mol g^{-1}) per unit time (s) produce the same angular rate in different devices. From Eq. 5, changing the current by the specific current, we obtain:

$$\omega\ (\mathrm{rad \cdot s^{-1}}) = k'\ (\mathrm{rad \cdot g \cdot s^{-1} \cdot A^{-1}}) * i\ (\mathrm{A \cdot g^{-1}}) \quad (6)$$

where the new constant, k', is the angle described when the reactive polymer consumes one unit of charge per gram of CP. This means that experiments from one muscle are only required in order to obtain this faradic characteristic of the CP. Then Eq. 6 describes the dynamics of any device prepared from the same material.

6.1.1 Prevailing anion exchange

Below are the results for a three-layer device consisting of pPy films, obtained using $LiClO_4$ as electrolyte, of dimensions 2 cm x 1.5 cm x 13 μm and 6 mg weight each. Such films have prevailing anion exchange. As anticipated, Fig. 4 shows the expected linear relationships between the current and the rate of the angular movement and electric charge and movement for a triple layer: the movement is under control of the current (Otero & Cortes 2004). Also a constant value of charge is consumed, irrespective of the amount of applied current, in agreement with the fact that the electrochemical reaction controls through the composition and the volume change of the polymeric films.

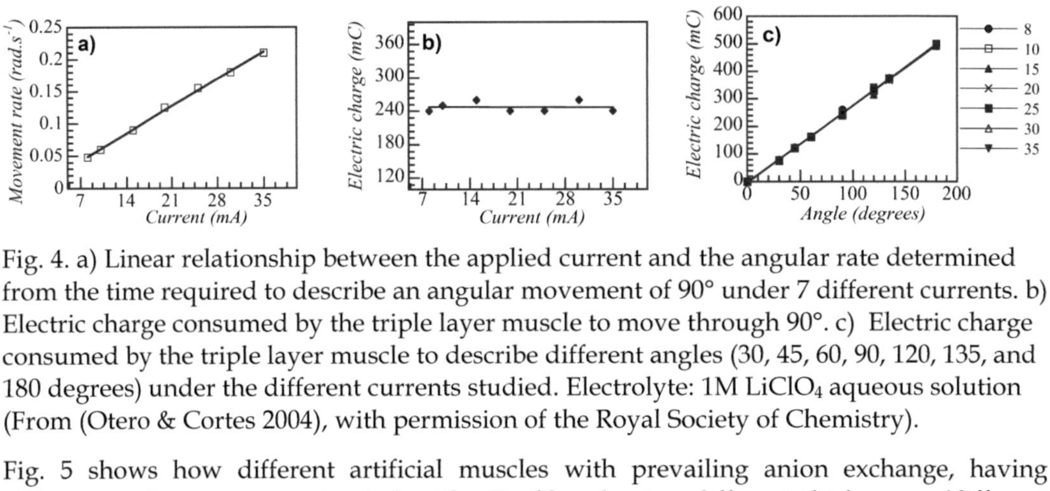

Fig. 4. a) Linear relationship between the applied current and the angular rate determined from the time required to describe an angular movement of 90° under 7 different currents. b) Electric charge consumed by the triple layer muscle to move through 90°. c) Electric charge consumed by the triple layer muscle to describe different angles (30, 45, 60, 90, 120, 135, and 180 degrees) under the different currents studied. Electrolyte: 1M $LiClO_4$ aqueous solution (From (Otero & Cortes 2004), with permission of the Royal Society of Chemistry).

Fig. 5 shows how different artificial muscles with prevailing anion exchange, having different surface area, constructed with pPy films having different thicknesses (different weight of pPy), produce the same angular movement rate under flow of analogous charge per unit time (current) and per unit weigh of CP (same variation of the oxidation).

6.1.2 Prevailing cation exchange

The linear relationships between electric charge and movement rate and applied current and movement rate are not exclusive of anion-exchange devices. Devices with prevailing cation exchange can also have the same properties. For example, a pPy-DBSA-ClO_4^-/non-conducting tape bilayer (Valero et al. 2011) also presents a linear relationship between angular rate and specific current for devices of different dimensions and weights, indicating that these are general properties of these materials, regardless of the type of material, device or species exchanged. (Fig. 6)

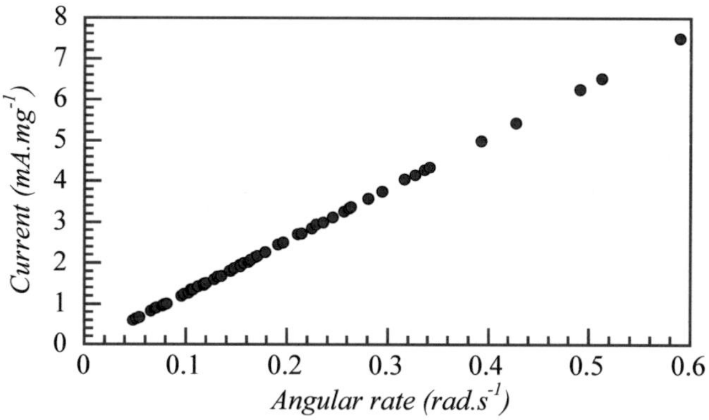

Fig. 5. Angular rate measured through a movement of 90 degrees using triple-layer muscle of polypyrrole (anion exchange) having different dimensions (including different weights of polypyrrole films: 8.3, 7.8, 7.4, 6, 5.5, 5.1 , 5, 4, 3.7, 3.5, 3, 2.3 and 2 mg) and checked in 1M $LiClO_4$ aqueous solution under different currents (10, 15, 20, 25 and 30 mA) (From (Otero & Cortes 2004), with permission of the Royal Society of Chemistry).

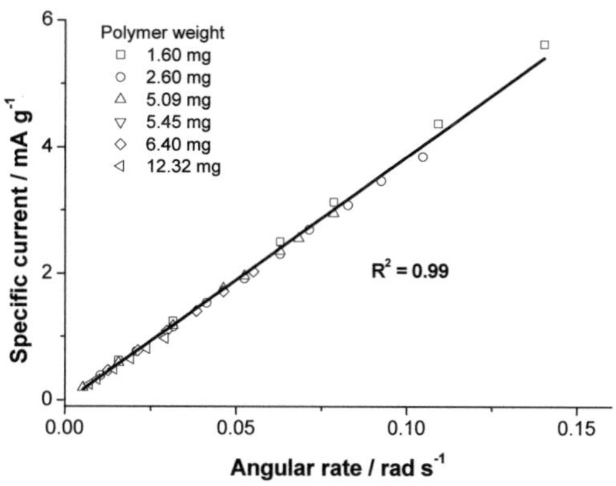

Fig. 6. Angular rate measured through a movement of π/4 radians using bilayer muscles of pPy/DBS (interchanging cations) having different dimensions (1.2x2.4; 1.0x1.5; 1.65x0.7, and 1.5x0.8 cm) with different weights, indicated on the figure. R^2 is the square of the correlation coefficient (From (Valero et al. 2011), with permission of Elsevier).

In conclusion, the faradic nature of the movement allows a perfect control of both the movement rate (by the current) and the angular position (by the charge). It means that we have a machine capable of transforming electrical energy into mechanical energy using molecular motors as mechanical components: the electrochemically stimulated conformational movements of the polymeric chains. The electrical generators nowadays available are able to give very precise electrical currents and soft current variations; therefore very precise movement rates and soft rate variations can be produced using artificial muscles, mimicking elegant movements from biological beings.

6.2 Other models

There are different methodologies that allow people working with artificial muscles to model their behaviour. Usually these methods are based on different aspects of the working principle of artificial muscles or they simply employ usual mathematical methods from different fields to develop the model.

6.2.1 Bending beam method

The bending beam method (Gere & Goodno 2009; Timoshenko 1925) is related to the mechanical study of solid state beam bending. The use of this model to explain the behaviour of artificial muscles is based on the study of the forces generated at the interface between the non-conductive layer, keeping its volume constant and the conducting polymer film, varying locally its volume. This model assumes that: (I) the thickness of the beam is small compared to the minimum radius of curvature, (II) there is a linear relationship between stress and strain of the material, and (III) Young's modulus, E, and the actuation coefficient of expansion of the conducting polymer, α, are constant and do not depend on spatial location inside each layer.

Pei and Inganäs (Pei & Inganäs 1993b; Pei & Inganäs 1993a; Pei & Inganäs 1992a) developed a model that correlates the local linear strain from a polypyrrole film and the curvature change suffered by an artificial muscle employing concepts from geometry and mechanics (Fig. 7). The actuator curvature radius (R_∞ is the radius at equilibrium and R_0 is the initial radius) is related to Young's modulus (E') and thicknesses (h) of the conducting and the non-conductive films (indicated by subscripts 1 and 2 respectively), and to the volume changes locally produced at the interface between both films $\alpha_{(t)}$, in Eq. 7.

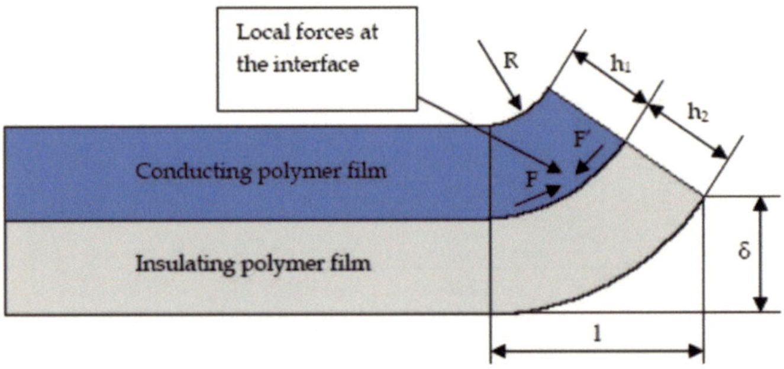

Fig. 7. Scheme of deformation during actuation used in the bending beam method.

$$\frac{1}{R_\infty} - \frac{1}{R_0} = \frac{6\alpha_{(t)}}{\dfrac{\left(E_1' h_1^2 - E_2' h_2^2\right)^2}{E_1' E_2' h_1 h_2 \left(h_1 + h_2\right)} + 4\left(h_1 + h_2\right)} \qquad (7)$$

Christophersen et al. (Christophersen et al. 2006) expanded the model by including strain and modulus variations along the direction of film thickness in order to obtain a tool to develop micro-actuators to achieve a specified radius of curvature. Also Alici et al. (Alici et al. 2006b) worked out in their approach, considering that the mechanical movements produced at a trilayer actuator were due to the charge transference occurring during redox reactions in conducting polymers. Actuator's position, rate of the movement and force generated by the actuator (Alici & Huynh 2006) were simulated. They applied this model to the study of a biomimetic device (propulsion fins) (Alici et al. 2007). Finally, Lee et al. (Lee et al. 2005) used this model for the movement of ionic polymer-metal composite actuators and more recently, Du et al. (Du et al. 2010) developed a general model for a multilayer system (N layers) to link the actuation strain of the actuator to the bending curvature.

6.2.2 Finite element model

Several authors have worked out models using the finite element method well known in engineering.

Alici et al. (Alici et al. 2006a; Metz et al. 2006) developed a model based on a lumped-parameter mathematical model for trilayer actuators employing the analogy between thermal strain and the real strain (due to the insertion/extraction of ions inside the polymeric film) in pPy actuators actuating in air, more difficult for simulation than thermal phenomena. The bending angle and the moment were solved by using finite element method and ANSYS software. An optimization of the geometry was performed as well, in order to obtain the greater output properties from a determined input voltage.

Shapiro and Smela (Shapiro & Smela 2007) developed a two dimensional model (along a full area) to obtain curvature and angular moment of bilayer and trilayer actuators (metal, polymeric buffer, non actuating but conducting, and polymeric actuator). Thus, they combined the results from the previous model (bending beam method) with finite element method to obtain the solution.

6.2.3 Equivalent transmission line model

This model is based on the development of electrical equivalent circuits giving similar electrical responses to that of the actuator. This resource is a very useful tool to develop models due to the great number of facilities available to study electrical circuits (they are easy to study in different steps or modules). Such models are employed in engineering and physics, or in electrochemistry in order to explain the capacitive behaviour of CP (Albery & Mount 1993; Albery & Mount 1994; Bisquert et al. 2000; Buck & Mundt 1999; Paasch 2000; Vorotyntsev et al. 1994).

Ren and Pickup (Ren & Pickup 1997) proposed equivalent electrical circuits to model the electron transport and electron transfer in composite pPy-PSS films based on the works of Albery. Also, Fang et al. (Fang et al. 2008; Yang et al. 2008) developed a scalable model (valid for different sizes of the actuator) including dynamic actuation performance under a given voltage input, joining three different modules that model three different aspects of the actuator: electrochemical dynamics, stress-generation relating transferred charges to internal actuation stress and mechanical dynamics. Finally, Shoa et al. (Shoa et al. 2011) developed a dynamic electromechanical model for electrochemically driven conducting polymer actuators based on a 2-D impedance model using an RC transmission line equivalent circuit to predict the charge transfer during actuation. Besides, a mechanical model (based on the bending beam model) is considered after the equivalent circuit that simulates ion "diffusion" through the thickness and electronic resistance along the length.

7. Simultaneous sensing capabilites

The electronic levels for the different compositions of those wet and reactive polymers are expected to respond very fast to the change of any external physical or chemical variable. These electronic levels are in fast equilibrium with electronic levels in the connecting metal, so the modification of the device potential, whatever maybe its origin, will be detected by the potentiostat simultaneously to the change of the variables. These fast events allow developing actuators able to sense the ambient conditions, during actuation (Otero 2008; Otero 2009). For example, the flow of a constant anodic current through a trilayer device under constant ambient conditions gives a progressive increase of the electrodic potential. (Otero & Cortes 2000; Otero & Cortes 2001).

Fig. 8 shows the evolution of device potential for different values of the surrounding variables: electrolyte concentration, temperature, different applied current or different weights attached to the bottom of the trilayer. As expected, rising electrolyte concentrations and rising working temperatures promote a shift of the device potential evolution while working, towards lower potentials. Rising weights attached to the device at the bottom require higher energies to keep a constant movement rate, under constant current, giving increasing device's potential. The flow of rising currents, assuming constant resistances (electrolyte, anodic and cathodic reactions), produces higher potentials and requires shorter times to flow the same charge describing the same angle.

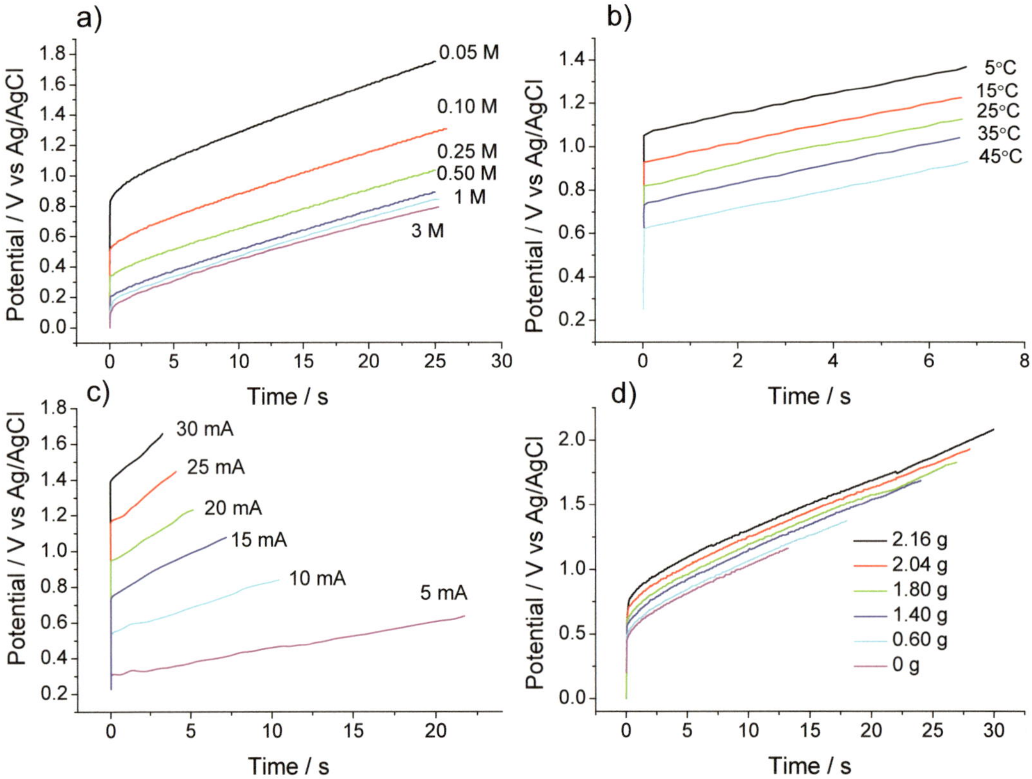

Fig. 8. Chronopotentiograms obtained when a triple layer (2 x 1.5 cm^2, 12 mg of pPy) describes 90° angular movenent a) in aqueous solutions of $LiClO_4$ (3, 1, 0.5, 0.25, 0.1, and 0.05 M) under a constant current of 10 mA, b) in 0.1 M $LiClO_4$ at different temperatures: 5, 15, 25, 35 and 45 °C. c) under flow of different currents: 5, 10, 15, 20, 25, and 30 mA and d) shifting different attached steel weights: 0.6, 1.4, 1.8, 2.04, and 2.16 g with a device of 12 mg of polymer weight (Adapted from (Otero & Cortes 2000; Otero & Cortes 2001), with permission).

Fig. 9 shows the linear evolution of the electrical energy consumed by the artificial muscle as a function of the studied experimental variables. Rising driving currents, or increasing shifted weights, produces higher overpotentials consuming greater electrical energy. Rising electrolyte concentrations, or temperatures, require lower overpotentials consuming lower electrical energies during the movement. Those results underline the simultaneous sensing capabilities of the device during the actuation process under reactive conditions and outside the chemical equilibrium. This is a general property for any device working by electrochemical reaction of conducting polymers such as batteries, smart windows, smart membranes, artificial muscles, electron-ion transducers, etc.

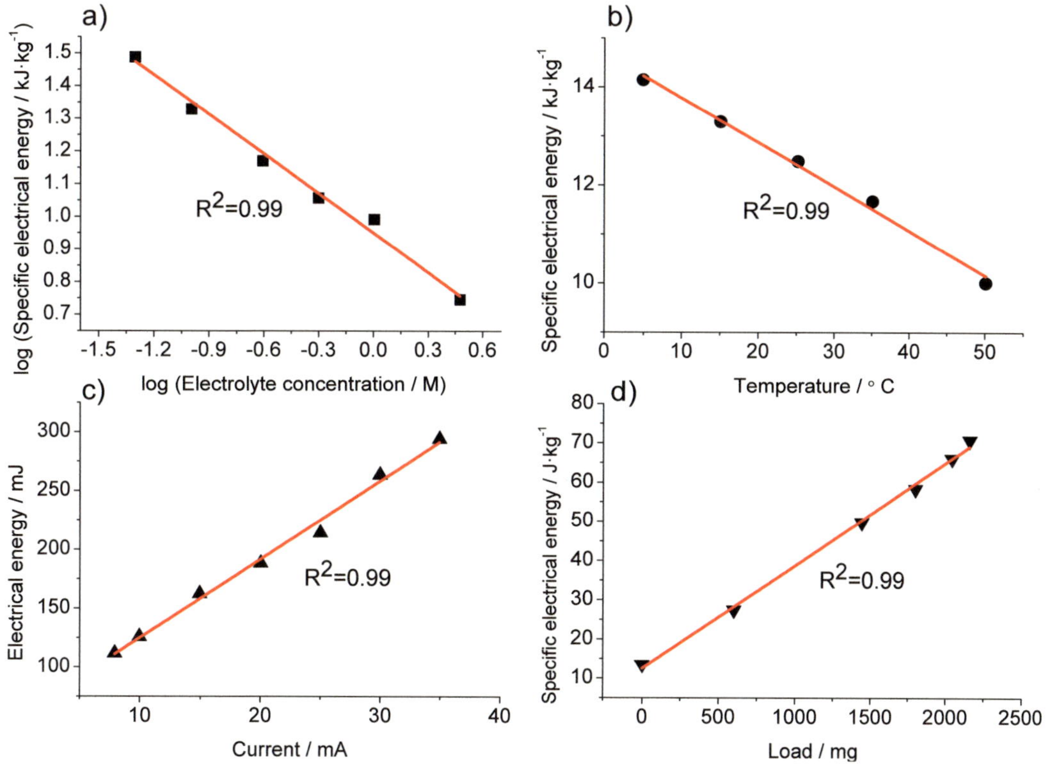

Fig. 9. Consumed electrical energy by the device in Fig. 8 as a function of the different studied variables: (a) electrolyte concentration, (b) temperature, (c) current, and (d) shifted weight (Adapted from (Otero & Cortes 2000; Otero & Cortes 2001), with permission).

8. Tactile sensitivity

By trailing increasing weights by the muscle we obtain increasing device potentials. If a trilayer or bilayer device moves freely and meets an obstacle, it must touch, push and shift the obstacle if possible. As observed in Fig. 10, at the beginning of the movement, the muscle moves freely until it touches the obstacle: the evolution of the muscle potential must be the same as that of the muscle moving freely, without any hanging weight or without any obstacle in its way. When the muscle touches the obstacle it feels a mechanical resistance and the muscle produces an extra energy drain by raising the muscle potential (Otero et al.

2004a; Otero & Cortes 2003; Otero et al. 2004b). The response is a potential step, proportional to the obstacle weight.

When the mechanical resistance of the obstacle exceeds the mechanical energy produced by the device, the muscle is unable to shift the obstacle and the muscle potential steps to very high values at the moment of the contact. The system indicates when a muscle, or the mechanical tool driven by the muscle, touches an obstacle how much mechanical resistance the obstacle opposes.

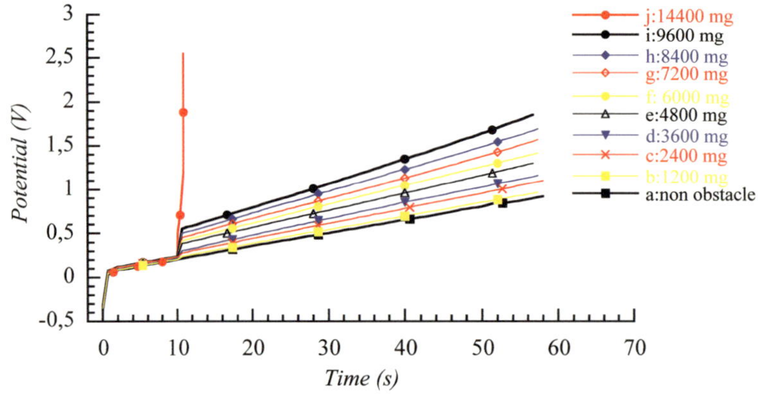

Fig. 10. Chronopotentiograms obtained from a triple layer device composed of two polypyrrole films [2 cm x 1.5 cm x 13 µm] weighing 6 mg each, under flow of 5 mA in 1 M $LiClO_4$. The muscle moves freely, meeting an obstacle after 10 s and sliding it for 3.5 s, overcomes its border and continues with a full angular movement. Obstacles weighing 1.2, 2.4, 3.6, 4.8, 6.0, 7.2, 8.4, 9.6 mg were slided, but the muscle was unable to push and slide an obstacle weighing 14.4 g (From (Otero & Cortes 2003), with permission of John Wiley & Sons).

9. Conclusions

Artificial muscles from reactive conducting polymers provide another unique property: simultaneous actuation and sensing. Both signals (driving current and sensing potential) are simultaneously incorporated in the two connecting wires. New scientific and technological challenges and borders are stated. Those borders require new theoretical models. Tentative electrochemical, mechanical, mathematical models are being developed for helping engineering designs and predictions. An integrated model is expected for those reactive dense and biomimetic materials and their applications in biomimetic devices.

10. Acknowledgment

Authors acknowledge financial support from Spanish Government (MCINN) Project MAT2008-06702, Seneca Foundation Project 08684/PI/08. J.A.P. thanks MCINN for Juan de la Cierva grant JCI-2008-02022. Cost action MP 1003 (ESNAM) from the EU.

11. References

Abidian, M. R., Kim, D. H. & Martin, D. C. (2006). Conducting-polymer nanotubes for controlled drug release, *Adv. Mater.*, Vol. 18, No. 4, pp. 405-409, ISSN 0935-9648.

Ahonen, H. J., Lukkari, J. & Kankare, J. (2000). n- and p-doped poly(3,4-ethylenedioxythiophene): Two electronically conducting states of the polymer, *Macromolecules*, Vol. 33, No. 18, pp. 6787-6793, ISSN 0024-9297.

Albery, W. J. & Mount, A. R. (1993). Application of A Transmission-Line Model to Impedance Studies on A Poly(Vinylferrocene)-Modified Electrode, *J.Chem.Soc.Faraday T.*, Vol. 89, No. 2, pp. 327-331, ISSN 0956-5000.

Albery, W. J. & Mount, A. R. (1994). Dual Transmission-Line with Charge-Transfer Resistance for Conducting Polymers, *J.Chem.Soc.Faraday T.*, Vol. 90, No. 8, pp. 1115-1119, ISSN 0956-5000.

Alici, G. & Huynh, N. N. (2006). Predicting force output of trilayer polymer actuators, *Sensor Actuat. A-Phys*, Vol. 132, No. 2, pp. 616-625, ISSN 0924-4247.

Alici, G., Metz, P. & Spinks, G. M. (2006a). A methodology towards geometry optimization of high performance polypyrrole (PPy) actuators, *Smart Mater.Struct.*, Vol. 15, No. 2, pp. 243-252, ISSN 0964-1726.

Alici, G., Mui, B. & Cook, C. (2006b). Bending modeling and its experimental verification for conducting polymer actuators dedicated to manipulation applications, *Sensor Actuat. A-Phys*, Vol. 126, No. 2, pp. 396-404, ISSN 0924-4247.

Alici, G., Spinks, G., Huynh, N. N., Sarmadi, L. & Minato, R. (2007). Establishment of a biomimetic device based on tri-layer polymer actuators-propulsion fins, *Bioinspir. Biomim.*, Vol. 2, No. 2, pp. S18-S30, ISSN 1748-3182.

Alici, G. & Gunderson, D. (2009). A bio-inspired robotic locomotion system based on conducting polymer actuators, In: *2009 IEEE/ASME International Conference on Advanced Intelligent Mechatronics, Vols 1-3*, pp. 998-1004, IEEE, ISBN 978-1-4244-2852-6, New York.

Arbizzani, C., Catellani, M., Mastragostino, M. & Mingazzini, C. (1995). N-doped and p-doped polydithieno[3,4-b-3',4'-d] thiophene - a narrow-band gap polymer for redox supercapacitors, *Electrochim. Acta*, Vol. 40, No. 12, pp. 1871-1876, ISSN 0013-4686.

Ariza, M. J. & Otero, T. F. (2007). Nitrate and chloride transport through a smart membrane, *J. Membr. Sci.*, Vol. 290, No. 1-2, pp. 241-249, ISSN 0376-7388.

Balzani, V., Credi, A., Ferrer, B., Silvi, S. & Venturi, M. (2005). *Artificial molecular motors and machines: Design principles and prototype systems*, Springer-Verlag, ISBN 0340-1022, Berlin.

Baughman, R. H. & Shacklette, L. W. (1991). Application of dopant-induced structure-property changes of conducting polymers, In: *Science and Application of Conducting Polymers*, W. R. Salaneck, D. T. Clark and E. J. Samuelsen, pp. 47-61, IOP Publishing Ltd, ISBN 0750300493, Bristol.

Baughman, R. H. (1996). Conducting polymer artificial muscles, *Synth. Met.*, Vol. 78, No. 3, pp. 339-353, ISSN 0379-6779.

Bisquert, J., Belmonte, G. G., Santiago, F. F., Ferriols, N., Yamashita, M. & Pereira, E. C. (2000). Application of a distributed impedance model in the analysis of conducting polymer films, *Electrochem.Commun.*, Vol. 2, No. 8, pp. 601-605, ISSN 1388-2481.

Blonsky, P. M. & Meridian, I. (1997). Structurally stable gelled electrolytes, US Patent 5648011.

Buck, R. P. & Mundt, C. (1999). Origins of finite transmission lines for exact representations of transport by the Nernst-Planck equations for each charge carrier, *Electrochim.Acta*, Vol. 44, No. 12, pp. 1999-2018, ISSN 0013-4686.

Chiang, C. K., Fincher, C. R., Park, Y. W., Heeger, A. J., Shirakawa, H., Louis, E. J., Gau, S. C. & Macdiarmid, A. G. (1977). Electrical-Conductivity in Doped Polyacetylene, *Phys. Rev. Lett.*, Vol. 39, No. 17, pp. 1098-1101, ISSN 0031-9007.

Cho, M., Seo, H., Nam, J., Choi, H., Koo, J. & Lee, Y. (2007). High ionic conductivity and mechanical strength of solid polymer electrolytes based on NBr/ionic liquid and its application to an electrochemical actuator, *Sensor Actuat. B-Chem*, Vol. 128, No. 1, pp. 70-74, ISSN 0925-4005.

Christophersen, M., Shapiro, B. & Smela, E. (2006). Characterization and modeling of PPy bilayer microactuators - Part 1. Curvature, *Sensor Actuat. B-Chem*, Vol. 115, No. 2, pp. 596-609, ISSN 0925-4005.

Davis, A. P. (1999). Nanotechnology - Synthetic molecular motors, *Nature*, Vol. 401, No. 6749, pp. 120-121, ISSN 0028-0836.

DellaSanta, A., DeRossi, D. & Mazzoldi, A. (1997a). Performance and work capacity of a polypyrrole conducting polymer linear actuator, *Synth. Met.*, Vol. 90, No. 2, pp. 93-100, ISSN 0379-6779.

DellaSanta, A., DeRossi, D. & Mazzoldi, A. (1997b). Characterization and modelling of a conducting polymer muscle-like linear actuator, *Smart. Mater. Struct.*, Vol. 6, No. 1, pp. 23-34, ISSN 0964-1726.

DellaSanta, A., Mazzoldi, A., Tonci, C. & De Rossi, D. (1997c). Passive mechanical properties of polypyrrole films: a continuum, poroelastic model, *Mater. Sci. Eng., C*, Vol. 5, No. 2, pp. 101-109, ISSN 0928-4931.

Deshpande, S. D., Kim, J. & Yun, S. R. (2005a). New electro-active paper actuator using conducting polypyrrole: actuation behaviour in LiClO4 acetonitrile solution, *Synth. Met.*, Vol. 149, No. 1, pp. 53-58, ISSN 0379-6779.

Deshpande, S. D., Kim, J. & Yun, S. R. (2005b). Studies on conducting polymer electroactive paper actuators: Effect of humidity and electrode thickness, *Smart Mater. Struct.*, Vol. 14, No. 4, pp. 876-880, ISSN 0964-1726.

Ding, J., Liu, L., Spinks, G. M., Zhou, D., Wallace, G. G. & Gillespie, J. (2003). High performance conducting polymer actuators utilising a tubular geometry and helical wire interconnects, *Synth. Met.*, Vol. 138, No. 3, pp. 391-398, ISSN 0379-6779.

Du, P., Lin, X. & Zhang, X. (2010). A multilayer bending model for conducting polymer actuators, *Sensor Actuat. A-Phys*, Vol. 163, No. 1, pp. 240-246, ISSN 0924-4247.

Dyer, A. L. & Reynolds, J. R. (2007). Electrochromism of Conjugated Conducting Polymers, In: *Handbook of Conducting Polymers Vol. 1*, T. A. Skotheim, R. L. Elsenbaumer and J. R. Reynolds, pp. 1-63 Ch. 20, CRC Press, ISBN 9781420043587, Boca Raton.

Fang, Y., Tan, X. O., Shen, Y. T., Xi, N. & Alici, G. (2008). A scalable model for trilayer conjugated polymer actuators and its experimental validation, *Mater. Sci. Eng., C*, Vol. 28, No. 3, pp. 421-428, ISSN 0928-4931.

Gere, J. M. & Goodno, B. J. (2009). *Mechanics of Materials 4th Ed.*, Cengage Learning, ISBN 0534553974, Toronto.

Han, G. & Shi, G. (2004a). Conducting polymer electrochemical actuator made of high-strength three-layered composite films of polythiophene and polypyrrole, *Sensor Actuat. B-Chem*, Vol. 99, No. 2-3, pp. 525-531, ISSN 0925-4005.

Han, G. & Shi, G. (2006). Electrochemical actuator based on single-layer polypyrrole film, *Sensor Actuat. B-Chem*, Vol. 113, No. 1, pp. 259-264, ISSN 0925-4005.

Han, G. Y. & Shi, G. Q. (2004b). High-response tri-layer electrochemical actuators based on conducting polymer films, *J. Electroanal. Chem.*, Vol. 569, No. 2, pp. 169-174, ISSN 0022-0728.

Hara, S., Zama, T., Sewa, S., Takashima, W. & Kaneto, K. (2003). Polypyrrole-metal coil composites as fibrous artificial muscles, *Chem. Lett.*, Vol. 32, No. 9, pp. 800-801, ISSN 0366-7022.

Hara, S., Zama, T., Ametani, A., Takashima, W. & Kaneto, K. (2004a). Enhancement in electrochemical strain of a polypyrrole-metal composite film actuator, *J. Mater. Chem.*, Vol. 14, No. 18, pp. 2724-2725, ISSN 0959-9428.

Hara, S., Zama, T., Takashima, W. & Kaneto, K. (2004b). Polypyrrole-metal coil composite actuators as artificial muscle fibres, *Synth. Met.*, Vol. 146, No. 1, pp. 47-55, ISSN 0379-6779.

He, X. M., Li, C., Chen, F. G. & Shi, G. Q. (2007). Polypyrrole microtubule actuators for seizing and transferring microparticles, *Adv. Funct. Mater.*, Vol. 17, No. 15, pp. 2911-2917, ISSN 1616-301X.

Heuer, H. W., Wehrmann, R. & Kirchmeyer, S. (2002). Electrochromic window based on conducting poly (3,4-ethylenedioxythiophene)poly(styrene sulfonate), *Adv. Funct. Mater.*, Vol. 12, No. 2, pp. 89-94, ISSN 1616-301X.

Higgins, S. J., Lovell, K. V., Rajapakse, R. M. G. & Walsby, N. M. (2003). Grafting and electrochemical characterisation of poly-(3,4-ethylenedioxythiophene) films, on Nafion and on radiation-grafted polystyrenesulfonate-polyvinylidene fluoride composite surfaces, *J. Mater. Chem.*, Vol. 13, No. 10, pp. 2485-2489, ISSN 0959-9428.

Hillman, A. R., Loveday, D. C., Swann, M. J., Eales, R. M., Hamnett, A., Higgins, S. J., Bruckenstein, S. & Wilde, C. P. (1989). Charge Transport in Electroactive Polymer-Films, *Faraday Discuss. Chem. Soc.*, Vol. 88, No., pp. 151-163, ISSN 0301-7249.

Huang, W. S., Humphrey, B. D. & MacDiarmid, A. G. (1986). Polyaniline, A Novel Conducting Polymer - Morphology and Chemistry of Its Oxidation and Reduction In Aqueous-Electrolytes, *J. Chem. Soc., Faraday Trans. I*, Vol. 82, No., pp. 2385-2400, ISSN 0300-9599.

Hutchison, A. S., Lewis, T. W., Moulton, S. E., Spinks, G. M. & Wallace, G. G. (2000). Development of polypyrrole-based electromechanical actuators, *Synth. Met.*, Vol. 113, No. 1-2, pp. 121-127, ISSN 0379-6779.

Ikushima, K., John, S., Yokoyama, K. & Nagamitsu, S. (2009). A practical multilayered conducting polymer actuator with scalable work output, *Smart. Mater. Struct.*, Vol. 18, No. 9, pp. 9, ISSN 0964-1726.

Inzelt, G. (2008). Redox Transformations and Transport Processes, In: *Conducting Polymers*, F. Scholz, pp. 169-224, Springer-Verlag Berlin, ISBN 9783540759294, Heidelberg (Germany).

Irvin, J. A., Irvin, D. J. & Stenger-Smith, J. D. (2007). Electroactive Polymers for Batteries and Supercapacitors, In: *Handbook of Conducting Polymers Vol. 2*, T. A. Skotheim, R. L. Elsenbaumer and J. R. Reynolds, pp. 9.1-9.29, CRC Press, ISBN 9781420043587, Boca Raton.

Isaksson, J., Tengstedt, C., Fahlman, M., Robinson, N. & Berggren, M. (2004). A solid-state organic electronic wettability switch, *Adv.Mater.*, Vol. 16, No. 4, pp. 316-320, ISSN 0935-9648.

Jager, E. W. H., Smela, E., Inganäs, O. & Lundstrom, I. (1999). Polypyrrole microactuators, *Synth. Met.*, Vol. 102, No. 1-3, pp. 1309-1310, ISSN 0379-6779.

Jager, E. W. H., Inganäs, O. & Lundstrom, I. (2000a). Microrobots for micrometer-size objects in aqueous media: Potential tools for single-cell manipulation, *Science*, Vol. 288, No. 5475, pp. 2335-2338, ISSN 0036-8075.

Jager, E. W. H., Smela, E. & Inganäs, O. (2000b). Microfabricating conjugated polymer actuators, *Science*, Vol. 290, No. 5496, pp. 1540-1545, ISSN 0036-8075.

Jager, E. W. H., Inganäs, O. & Lundstrom, I. (2001). Perpendicular actuation with individually controlled polymer microactuators, *Adv. Mater.*, Vol. 13, No. 1, pp. 76-79, ISSN 0935-9648.

Jager, E. W. H., Immerstrand, C., Peterson, K. H., Magnusson, K.-E., Lundström, I. & Inganäs, O. (2002). The Cell Clinic: Closable Microvials for Single Cell Studies, *Biomed. Microdev*, Vol. 4, No. 3, pp. 177-187, ISSN 1387-2176.

John, S., Alici, G. & Cook, C. (2008). Frequency response of polypyrrole trilayer actuator displacement, *Proceedings of Electroactive Polymer Actuators and Devices (Eapad) 2008*, pp. T9271, ISBN 0277-786X, San Diego (USA).

Kaneto, K., Sonoda, Y. & Takashima, W. (2000). Direct measurement and mechanism of electro-chemomechanical expansion and contraction in polypyrrole films, *Jpn. J. Appl. Phys. 1*, Vol. 39, No. 10, pp. 5918-5922, ISSN 0021-4922.

Kiefer, R., Mandviwalla, X., Archer, R., Tjahyono, S. S., Wang, H., MacDonald, B., Bowmaker, G. A., Kilmartin, P. A. & Travas-Sejdic, J. (2008). The application of polypyrrole trilayer actuators in microfluidics and robotics, *Proceedings of Electroactive Polymer Actuators and Devices (Eapad) 2008*, pp. E9271, ISBN 0277-786X.

Lee, K. K. C., Munce, N. R., Shoa, T., Charron, L. G., Wright, G. A., Madden, J. D. & Yang, V. X. D. (2009). Fabrication and characterization of laser-micromachined polypyrrole-based artificial muscle actuated catheters, *Sensor Actuat. A-Phys*, Vol. 153, No. 2, pp. 230-236, ISSN 0924-4247.

Lee, S., Park, H. C. & Kim, K. J. (2005). Equivalent modeling for ionic polymer-metal composite actuators based on beam theories, *Smart Mater.Struct.*, Vol. 14, No. 6, pp. 1363-1368, ISSN 0964-1726.

Li, W. G., Johnson, C. L. & Wang, H. L. (2004). Preparation and characterization of monolithic polyaniline-graphite composite actuators, *Polymer*, Vol. 45, No. 14, pp. 4769-4775, ISSN 0032-3861.

Lu, W., Fadeev, A. G., Qi, B. H., Smela, E., Mattes, B. R., Ding, J., Spinks, G. M., Mazurkiewicz, J., Zhou, D. Z., Wallace, G. G., MacFarlane, D. R., Forsyth, S. A. & Forsyth, M. (2002). Use of ionic liquids for pi-conjugated polymer electrochemical devices, *Science*, Vol. 297, No. 5583, pp. 983-987, ISSN 0036-8075.

Ludwig, K. A., Uram, J. D., Yang, J. Y., Martin, D. C. & Kipke, D. R. (2006). Chronic neural recordings using silicon microelectrode arrays electrochemically deposited with a poly(3,4-ethylenedioxythiophene) (PEDOT) film, *J. Neural Eng.*, Vol. 3, No. 1, pp. 59-70, ISSN 1741-2560.

Mazzoldi, A., Degl'Innocenti, C., Michelucci, M. & De Rossi, D. (1998). Actuative properties of polyaniline fibers under electrochemical stimulation, *Mater. Sci. Eng., C*, Vol. 6, No. 1, pp. 65-72, ISSN 0928-4931.

Metz, P., Alici, G. & Spinks, G. M. (2006). A finite element model for bending behaviour of conducting polymer electromechanical actuators, *Sensor Actuat. A-Phys*, Vol. 130, No., pp. 1-11, ISSN 0924-4247.

Naka, Y., Fuchiwaki, M. & Tanaka, K. A micropump driven by a polypyrrole-based conducting polymer soft actuator, *Polym. Int.*, Vol. 59, No. 3, pp. 352-356, ISSN 1097-0126.

Nakano, T. & Okamoto, Y. (2001). Synthetic Helical Polymers: Conformation and Function, *Chem. Rev.*, Vol. 101, No. 12, pp. 4013-4038, ISSN

Okamoto, T., Tada, K. & Onoda, M. (2000). Bending machine using anisotropic polypyrrole films, *Jpn. J. Appl. Phys. 1*, Vol. 39, No. 5A, pp. 2854-2858, ISSN 0021-4922.

Okamoto, T., Kato, Y., Tada, K. & Onoda, M. (2001). Actuator based on doping/undoping-induced volume change in anisotropic polypyrrole film, *Thin Solid Films*, Vol. 393, No. 1-2, pp. 383-387, ISSN 0040-6090.

Okuzaki, H. & Hattori, T. (2003). Electrically induced anisotropic contraction of polypyrrole films, *Synth. Met.*, Vol. 135-136, No., pp. 45-46, ISSN 0379-6779.

Okuzaki, H. (2008). A biomorphic origami actuator fabricated by folding a conducting paper, *J. Phys. Conf. Ser.*, Vol. 127, No. 1, pp. 012001, ISSN 1742-6596.

Onoda, M., Okamoto, T., Tada, K. & Nakayama, H. (1999a). Polypyrrole films with anisotropy for artificial muscles and examination of bending behavior, *Jpn. J. Appl. Phys. 2*, Vol. 38, No. 9AB, pp. L1070-L1072, ISSN 0021-4922.

Onoda, M., Tada, K. & Nakayama, H. (1999b). Polypyrrole films with anisotropy, *Synth. Met.*, Vol. 102, No. 1-3, pp. 1321-1322, ISSN 0379-6779.

Onoda, M. & Tada, K. (2004). Anisotropic bending machine using conducting polypyrrole, *IEICE Trans. Electron.*, Vol. E87C, No. 2, pp. 128-135, ISSN 0916-8524.

Onoda, M., Shonaka, H. & Tada, K. (2005). A self-organized bending-beam electrochemical actuator, *Curr. Appl. Phys.*, Vol. 5, No. 2, pp. 194-201, ISSN 1567-1739.

Orata, D. & Buttry, D. A. (1987). Determination of ion populations and solvent content as functions of redox state and pH in polyaniline, *J. Am. Chem. Soc.*, Vol. 109, No. 12, pp. 3574-3581, ISSN 0002-786.

Otero, T. F., Angulo, E., Rodriguez, J. & Santamaria, C. (1992a). Electrochemomechanical Properties from A Bilayer - Polypyrrole Nonconducting and Flexible Material Artificial Muscle, *J. Electroanal. Chem.*, Vol. 341, No. 1-2, pp. 369-375, ISSN 0022-0728.

Otero, T. F., Angulo, E., Rodriguez, J. & Santamaria, C. (1992b). Dispositivos laminares que emplean polímeros conductores capaces de provocar movimientos mecánicos, ES Patent 2 048 086.

Otero, T. F., Rodriguez, J. & Santamaria, C. (1992c). Músculos artificiales formados por multicapas: polímeros conductores-no conductores, ES Patent 2 062 930.

Otero, T. F. (1999). Conducting polymers, electrochemistry, and biomimicking processes, In: *Modern Aspects of Electrochemistry*, J. O. M. Bockris, R. E. White and B. E. Conway, pp. 307-434, Kluwer Academic/Plenum Publ, ISBN 0076-9924, New York.

Otero, T. F. (2000). Biomimicking materials with smart polymers, In: *Structural biological materials. Design and structure-properties relationships*, M. Elices and R. W. Cahn, pp. 187-220, Pergamon Materials Series, ISBN 0080434169, Amsterdam (The Netherlands).

Otero, T. F. & Cortes, M. T. (2000). Electrochemical characterization and control triple-layer muscles, *Proceedings of Smart Structures and Materials 2000: Electroactive Polymer Actuators and Devices (Eapad)*, pp. 252-260, ISBN 0277-786X, Newport Beach, USA.

Otero, T. F. & Cortes, M. T. (2001). Characterization of triple layers, *Proceedings of Smart Structures and Materials 2001: Electroactive Polymer Actuators and Devices*, pp. 93-100, ISBN 0277-786X, Newport Beach, USA.

Otero, T. F., Cortes, M. T. & Boyano, I. (2002). Macroscopic devices and complex movements developed with artificial muscles, *Proceedings of Smart Structures and Materials 2002: Electroactive Polymer Actuators and Devices (Eapad)*, pp. 395-402, ISBN 0277-786X, San Diego, USA

Otero, T. F. & Cortes, M. T. (2003). Artificial muscles with tactile sensitivity, *Adv. Mater.*, Vol. 15, No. 4, pp. 279-282, ISSN 0935-9648.

Otero, T. F., Boyano, I., Cortes, M. T. & Vazquez, G. (2004a). Nucleation, non-stoiquiometry and sensing muscles from conducting polymers, *Electrochim. Acta*, Vol. 49, No. 22-23, pp. 3719-3726, ISSN 0013-4686.

Otero, T. F. & Cortes, M. T. (2004). Artificial muscle: movement and position control, *Chem. Commun.*, Vol., No. 3, pp. 284-285, ISSN 1359-7345.

Otero, T. F., Cortes, M. T., Boyano, L. & Vazquez, G. (2004b). Nucleation, non-stoiquiometry, and tactile muscles with conducting polymers, *Proceedings of Smart Structures and Materials 2004: Electroactive Polymer Actuators and Devices (Eapad)*, pp. 425-432, ISBN 0277-786X, San Diego, USA.

Otero, T. F. & Broschart, M. (2006). Polypyrrole artificial muscles: a new rhombic element. Construction and electrochemomechanical characterization, *J. Appl. Electrochem.*, Vol. 36, No. 2, pp. 205-214, ISSN 0021-891X.

Otero, T. F., Cortes, M. T. & Arenas, G. V. (2007). Linear movements from two bending triple-layers, *Electrochim. Acta*, Vol. 53, No. 3, pp. 1252-1258, ISSN 0013-4686.

Otero, T. F. (2008). Artificial Muscles, Sensing and Multifunctionality, In: *Intelligent Materials*, M. Shahinpoor and H.-J. Schenider, pp. 142-190, Royal Society of Chemistry, ISBN 9780854043354, Cambridge (U.K.).

Otero, T. F. (2009). Soft, wet, and reactive polymers. Sensing artificial muscles and conformational energy, *J. Mater. Chem.*, Vol. 19, No., pp. 681-689, ISSN 0959-9428.

Paasch, G. (2000). The transmission line equivalent circuit model in solid-state electrochemistry, *Electrochem.Commun.*, Vol. 2, No. 5, pp. 371-375, ISSN 1388-2481.

Pandey, S. S., Takashima, W. & Kaneto, K. (2003). Structure property correlation: electrochemomechanical deformation in polypyrrole films, *Thin Solid Films*, Vol. 438, No., pp. 206-211, ISSN 0040-6090.

Pede, D., Smela, E., Johansson, T., Johansson, M. & Inganäs, O. (1998). A general-purpose conjugated-polymer device array for imaging, *Adv. Mater.*, Vol. 10, No. 3, pp. 233-237, ISSN 0935-9648.

Pei, Q. & Inganäs, O. (1993a). Electrochemical applications of the bending beam method; a novel way to study ion transport in electroactive polymers, *Solid State Ionics*, Vol. 60, No. 1-3, pp. 161-166, ISSN 0167-2738.

Pei, Q. & Inganäs, O. (1993b). Electrochemical applications of the bending beam method. 2. Electroshrinking and slow relaxation in polypyrrole, *J. Phys. Chem.*, Vol. 97, No. 22, pp. 6034-6041, ISSN 0022-3654.

Pei, Q. B. & Inganäs, O. (1992a). Electrochemical Applications of the Bending Beam Method .1. Mass-Transport and Volume Changes in Polypyrrole During Redox, *J. Phys. Chem.*, Vol. 96, No. 25, pp. 10507-10514, ISSN 0022-3654.

Pei, Q. B. & Inganäs, O. (1992b). Conjugated Polymers and the Bending Cantilever Method - Electrical Muscles and Smart Devices, *Adv. Mater.*, Vol. 4, No. 4, pp. 277-278, ISSN 0935-9648.

Pellegrino, J. (2003). The Use of Conducting Polymers in Membrane-Based Separations, *Ann. N. Y. Acad. Sci.*, Vol. 984, No. 1, pp. 289-305, ISSN 1749-6632.

Plesse, C., Vidal, F., Randriamahazaka, H., Teyssie, D. & Chevrot, C. (2005). Synthesis and characterization of conducting interpenetrating polymer networks for new actuators, *Polymer*, Vol. 46, No. 18, pp. 7771-7778, ISSN 0032-3861.

Plesse, C., Vidal, F., Teyssie, D. & Chevrot, C. (2010). Conducting polymer artificial muscle fibres: toward an open air linear actuation, *Chem. Commun.*, Vol. 46, No. 17, pp. 2910-2912, ISSN 1359-7345.

Qi, B. H., Lu, W. & Mattes, B. R. (2004). Strain and energy efficiency of polyaniline fiber electrochemical actuators in aqueous electrolytes, *J. Phys. Chem. B*, Vol. 108, No. 20, pp. 6222-6227, ISSN 1520-6106.

Ramírez-García, S. & Diamond, D. (2007). Biomimetic, low power pumps based on soft actuators, *Sensor Actuat. A-Phys*, Vol. 135, No. 1, pp. 229-235, ISSN 0924-4247.

Ranger, M. & Leclerc, M. (1998). Optical and electrical properties of fluorene-based pi-conjugated polymers, *Can. J. Chem.*, Vol. 76, No. 11, pp. 1571-1577, ISSN 0008-4042.

Ren, X. & Pickup, P. G. (1997). An impedance study of electron transport and electron transfer in composite polypyrrole + polystyrenesulphonate films, *J.Electroanal.Chem.*, Vol. 420, No. 1-2, pp. 251-257, ISSN 1572-6657.

Roemer, M., Kurzenknabe, T., Oesterschulze, E. & Nicoloso, N. (2002). Microactuators based on conducting polymers, *Anal. Bioanal. Chem.*, Vol. 373, No. 8, pp. 754-757, ISSN 1618-2642.

Sansinena, J. M., Olazabal, V., Otero, T. F., daFonseca, C. N. P. & DePaoli, M. A. (1997). A solid state artificial muscle based on polypyrrole and a solid polymeric electrolyte working in air, *Chem. Commun.*, Vol., No. 22, pp. 2217-2218, ISSN 1359-7345.

Sansiñena, J. M., Gao, J. B. & Wang, H. L. (2003). High-performance, monolithic polyaniline electrochemical actuators, *Adv. Funct. Mater.*, Vol. 13, No. 9, pp. 703-709, ISSN 1616-301X.

Santa, A. D., Mazzoldi, A. & de Rossi, D. (1996). Steerable Microcatheters Actuated by Embedded Conducting Polymer Structures, *J. Intel. Mat. Syst. Str.*, Vol. 7, No. 3, pp. 292-300, ISSN 1045-389X.

Schmidt, C. E., Shastri, V. R., Vacanti, J. P. & Langer, R. (1997). Stimulation of neurite outgrowth using an electrically conducting polymer, *Proc. Natl. Acad. Sci. U. S. A.*, Vol. 94, No. 17, pp. 8948-8953, ISSN 0027-8424.

Shakuda, S., Morita, S., Kawai, T. & Yoshino, K. (1993). Dynamic Characteristics of Bimorph with Conducting Polymer Gel, *Jpn. J. Appl. Phys. 1*, Vol. 32, No. 11A, pp. 5143-5146, ISSN 0021-4922.

Shapiro, B. & Smela, E. (2007). Bending Actuators with Maximum Curvature and Force and Zero Interfacial Stress, *J. Intel. Mat. Syst. Str.*, Vol. 18, No. 2, pp. 181-186, ISSN 1045-389X.

Shirakawa, H., Louis, E. J., Macdiarmid, A. G., Chiang, C. K. & Heeger, A. J. (1977). Synthesis of Electrically Conducting Organic Polymers - Halogen Derivatives of Polyacetylene, (Ch)X, *J. Chem. Soc., Chem. Commun.*, Vol., No. 16, pp. 578-580, ISSN 0022-4936.

Shoa, T., Madden, J. D., Fekri, N., Munce, N. R. & Yang, V. X. D. (2008). Conducting polymer based active catheter for minimally invasive interventions inside arteries, *Proceedings of Engineering in Medicine and Biology Society, 2008. EMBS 2008. 30th Annual International Conference of the IEEE*, pp. 2063-2066, ISBN 1557-170X.

Shoa, T., Yoo, D. S., Walus, K. & Madden, J. D. W. (2011). A Dynamic Electromechanical Model for Electrochemically Driven Conducting Polymer Actuators, *IEEE ASME Trans. Mechatronics*, Vol. 16, No. 1, pp. 42-49, ISSN 1083-4435.

Smela, E., Inganäs, O., Pei, Q. B. & Lundstrom, I. (1993). Electrochemical Muscles - Micromachining Fingers and Corkscrews, *Adv. Mater.*, Vol. 5, No. 9, pp. 630-632, ISSN 0935-9648.

Smela, E., Inganäs, O. & Lundstrom, I. (1995). Controlled Folding of Micrometer-Size Structures, *Science*, Vol. 268, No. 5218, pp. 1735-1738, ISSN 0036-8075.

Smela, E. (1999). A microfabricated movable electrochromic "pixel" based on polypyrrole, *Adv. Mater.*, Vol. 11, No. 16, pp. 1343-1345, ISSN 0935-9648.

Smela, E. & Gadegaard, N. (1999). Surprising volume change in PPy(DBS): An atomic force microscopy study, *Adv. Mater.*, Vol. 11, No. 11, pp. 953-957, ISSN 0935-9648.

Smela, E., Christophersen, M., Prakash, S. B., Urdaneta, M., Dandin, M. & Abshire, P. (2007). Integrated cell-based sensors and "cell clinics" utilizing conjugated polymer actuators, *Proceedings of Proc. SPIE-Int. Soc. Opt. Eng.*, pp., ISBN.

Snook, G. A., Kao, P. & Best, A. S. (2011). Conducting-polymer-based supercapacitor devices and electrodes, *J. Power Sources*, Vol. 196, No. 1, pp. 1-12, ISSN 0378-7753.

Song, M. K., Cho, J. Y., Cho, B. W. & Rhee, H. W. (2002). Characterization of UV-cured gel polymer electrolytes for rechargeable lithium batteries, *J. Power Sources*, Vol. 110, No. 1, pp. 209-215, ISSN 0378-7753.

Spinks, G. M., Wallace, G. G., Ding, J., Zhou, D., Xi, B. & Gillespie, J. (2003a). Ionic liquids and polypyrrole helix tubes: bringing the electronic Braille screen closer to reality, *Proceedings of Smart Structures and Materials 2003: Electroactive Polymer Actuators and Devices (EAPAD)*, pp. 372-380, ISBN 9780819448569, San Diego, USA.

Spinks, G. M., Zhou, D. Z., Liu, L. & Wallace, G. G. (2003b). The amounts per cycle of polypyrrole electromechanical actuators, *Smart. Mater. Struct.*, Vol. 12, No. 3, pp. 468-472, ISSN 0964-1726.

Svirskis, D., Wright, B. E., Travas-Sejdic, J., Rodgers, A. & Garg, S. (2010). Evaluation of physical properties and performance over time of an actuating polypyrrole based drug delivery system, *Sensor Actuat. B-Chem*, Vol. 151, No. 1, pp. 97-102, ISSN 0925-4005.

Takashima, W., Uesugi, T., Fukui, M., Kaneko, M. & Kaneto, K. (1997). Mechanochemoelectrical effect of polyaniline film, *Synth. Met.*, Vol. 85, No. 1-3, pp. 1395-1396, ISSN 0379-6779.

Takashima, W., Pandey, S. S. & Kaneto, K. (2003a). Investigation of bi-ionic contribution for the enhancement of bending actuation in polypyrrole film, *Sens. Actuators, B*, Vol. 89, No. 1-2, pp. 48-52, ISSN 0925-4005.

Takashima, W., Pandey, S. S. & Kaneto, K. (2003b). Bi-ionic actuator by polypyrrole films, *Synth. Met.*, Vol. 135, No. 1-3, pp. 61-62, ISSN 0379-6779.

Timoshenko, S. (1925). Analysis of bi-metal thermostats, *J. Opt. Soc. Am.*, Vol. 11, No. 3, pp. 233-255, ISSN 0093-4119.

Torresi, R. M. & Maranhao, S. L. D. (1999). Anion and solvent exchange as a function of the redox states in polyaniline films, *J. Electrochem. Soc.*, Vol. 146, No. 11, pp. 4179-4182, ISSN 0013-4651.

Tsai, E. W., Pajkossy, T., Rajeshwar, K. & Reynolds, J. R. (1988). Anion-Exchange Behavior of Polypyrrole Membranes, *J. Phys. Chem.*, Vol. 92, No. 12, pp. 3560-3565, ISSN 0022-3654.

Valero, L., Arias-Pardilla, J., Cauich-Rodríguez, J., Smit, M. A. & Otero, T. F. (2011). Characterization of the movement of polypyrrole-dodecylbenzenesulfonate-perchlorate/tape artificial muscles. Faradaic control of reactive artificial molecular motors and muscles, *Electrochim. Acta*, Vol. 56, No. 10, pp. 3721-3726, ISSN 0013-4686.

Vidal, F., Popp, J. F., Plesse, C., Chevrot, C. & Teyssie, D. (2003). Feasibility of conducting semi-interpenetrating networks based on a poly(ethylene oxide) network and poly(3,4-ethylenedioxythiophene) in actuator design, *J. Appl. Polym. Sci.*, Vol. 90, No. 13, pp. 3569-3577, ISSN 0021-8995.

Vidal, F., Plesse, C., Palaprat, G., Juger, J., Citerin, J., Kheddar, A., Chevrot, C. & Teyssie, D. (2009). Synthesis and Characterization of IPNs for Electrochemical Actuators, In: *Artificial Muscle Actuators Using Electroactive Polymers*, P. Vincenzini, Y. BarCohen and F. Carpi, pp. 8-17, Trans Tech Publications Ltd, ISBN 978-3-908158-27-1, Stafa-Zurich.

Vorotyntsev, M. A., Daikhin, L. I. & Levi, M. D. (1994). Modeling the Impedance Properties of Electrodes Coated with Electroactive Polymer-Films, *J.Electroanal.Chem.*, Vol. 364, No. 1-2, pp. 37-49, ISSN 0022-0728.

Wang, H. L., Gao, J. B., Sansiñena, J. M. & McCarthy, P. (2002). Fabrication and characterization of polyaniline monolithic actuators based on a novel configuration: Integrally skinned asymmetric membrane, *Chem. Mater.*, Vol. 14, No. 6, pp. 2546-2552, ISSN 0897-4756.

Wu, Y., Zhou, D., Spinks, G. M., Innis, P. C., Megill, W. M. & Wallace, G. G. (2005). TITAN: a conducting polymer based microfluidic pump, *Smart. Mater. Struct.*, Vol. 14, No. 6, pp. 1511-1516, ISSN 0964-1726.

Yang, F., Xiaobo, T. & Alici, G. (2008). Robust Adaptive Control of Conjugated Polymer Actuators, *IEEE Trans. Control Syst. Technol.*, Vol. 16, No. 4, pp. 600-612, ISSN 1063-6536.

Yao, Q., Alici, G. & Spinks, G. A. (2008). Feedback control of tri-layer polymer actuators to improve their positioning ability and speed of response, *Sensor Actuat. A-Phys*, Vol. 144, No. 1, pp. 176-184, ISSN 0924-4247.

Zama, T., Hara, S., Takashima, W. & Kaneto, K. (2004). The correlation between electrically induced stress and mechanical tensile strength of polypyrrole actuators, *Bull. Chem. Soc. Jpn.*, Vol. 77, No. 7, pp. 1425-1426, ISSN 0009-2673.

Part 4

Nanotechnology

6

Comparative Studies of Chemically Synthesized Polymers Aniline and o-Toluidine Nanocomposite Using Algerian Montmorillonite

Benyoucef Adelghani[1], Yahiaoui Ahmed[1], Hachemaoui Aicha[1], Sanchís Carlos[2] and Morallon Emilia[2]
[1]*Laboratoire de chimie organique, macromoléculaire et des matériaux, Université de Mascara,*
[2]*Departamento de Química Física e Instituto Universitario de Materiales, Universidad de Alicante*
[1]*Algeria*
[2]*Spain*

1. Introduction

Electronically conducting polymers have been the subject of numerous investigations in the past two decades (Nalwa, 1997a; Novak et al., 1997; Miller et al., 1997; Skotheim et al., 1998). These materials have properties that make them suitable for several applications including light-emitting diodes, sensors, batteries, and electrochemical supercapacitors. For the majority of these studies, a single polymer was prepared by chemical or electrochemical oxidation of the corresponding monomer. This as grown oxidized polymer is not pure and can be actually considered as a composite since counterions are included in the polymer matrix as dopant. On the other hand, the formation of copolymer displaying electronic conductivity has been much less investigated (Gningue et al., 1988; Laborde et al., 1990; Peters et al., 1992; Talu et al., 1996; Nalwa, 1997b; Sanchez De Pinto et al., 1997; Cha, 1997; San et al., 1998; Khalkali, 2005; Yildiz et al., 2006). The main motivation for preparing copolymer composites lies in the possibility that these materials will display better properties and also to overcome the limitation of the rareness of new conjugated ð-bond-containing monomers.

The preparation of copolymers from a pair of monomers will lead to an increase of the number of conductive polymers obtained from the same set of monomers (San et al., 1998). However, the main drawback of PANI is the limited pH range in which PANI retains its electrochemical activity. This is to some extend a function of the counter ion balancing the positive charge of the protonated emeraldine base (Lindfors et al., 2004, 2008). But usually PANI becomes insulating and electrochqemically inactive at pH above 4 (Diaz et al., 1980), which limits its use as a sensing platform for applications requiring neutral or alkaline electrolytes. Many approaches have been used to overcome this problem. These are based on (i) post-polymerization grafting of sulfonic groups onto PANI chains by fuming sulfuric acid treatment (Li et al., 2005), copolymerization of parent aniline with -COOH or -SO_3H

functionalized aniline derivatives (Karyakin et al., 1994; Xu et al., 1997; Barbero et al., 2004; Benyoucef et al., 2010), or performing electropolymerization in the presence of various organic acids (Sun et al., 1998; Mu et al., 2002, 2003; Zhang et al., 2004, 2007; Barrios et al., 2006; Blomquist et al., 2009).

All these approaches, by introducing an ionogenic acidic group to the PANI structure, hinder the deprotonation of the conducting form of PANI (emeraldine) and thus extend its electroactivity toward less acidic pHs.

At the initial stages of the electrochemical polymerization, aniline is oxidized and becomes a radical cation, which leads to dimerization. The dimers of aniline, p-aminodiphenylamine (ADPA) as a dominant product along with benzidine and hydrazobenzene as minor products, are immediately oxidized and conjugated with aniline monomer. Polyaniline is formed on the electrode surface as a film by repeating this electrochemical process. However, the actual overall reaction seems to be more complicated, and the properties of the polymer formed by this method are changeable with the reaction conditions, such as applied potential, aniline concentration, pH, and electrode material.

With regard to polyaniline based copolymers, a pioneering work has been done by Borole et al. (2004) and Wei et al. (1989, 1990). They have reported that aniline could be copolymerized with o-toludine to give rise to a copolymer film, of which conductivity could be controlled in a broad range. Copolymerization of aniline with N-butylaniline has been also reported (Bergeron et al., 1991). Copolymerization of aniline and o-toluidine by electrochemical and chemical method has been also studied (Dhawan et al., 1993). The resulting copolymer film found to have both good conductivity and solubility in common organic solvents. Lately, copolymerization of aniline with N-methylaniline (Langer et al., 1993) and that with 3-aminophenyl-boric acid (Porter et al., 1990) have also been documented and also the electrochemical copolymerization of aniline with o-aminobenzonitrile has been studied (Sato et al., 1994). These reports suggest that the copolymerization could provide us a convenient synthetic method to prepare new conducting materials with desired properties.

Here, we reported a preliminary spectroelectrochemical study of conducting polymers created using non-toxic cationic catalyst, known-as Magnhite-H^+ (Mag-H) (Yang et al., 1990; Belbachir et al., 2001). The polymerisation has been performed with aniline "polyaniline" and it has been extended to o-toluidine "poly(o-toluidine)" and copolymers of them. The properties of these polymers have been compared with those synthesized by bulk polymerisation. The relative reactivities are analyzed and the copolymers are characterized. The effect of the copolymer composition on polymer properties such as conductivity, UV-visible (UV-vis) and FT-IR spectroscopies, X-ray diffraction (XRD), Transmission electron micrographs (TEM), X-ray photoelectronic spectroscopy (XPS) and electrochemical response were evaluated.

2. Experimental

2.1 Materials

Aniline (from Aldrich) was distilled under vacuum prior to use and o-toluidine (from Aldrich) was used as received. Perchloric acid (from Merck) was suprapur quality and the water employed for the preparation of the solutions was obtained from an Elga Labwater Purelab Ultra system. A natural sodium montmorillonite clay (named as Maghnite) obtained from Tlemcen (Algeria) has been used

2.2 Catalyst structure

Various methods of analysis, such as 27Al and 29Si MAS NMR, show that Maghnite is a montmorillonite silicate. The elementary analysis of the selected samples obtained using X-Ray Fluorescence (XRF) is presented in Table 1. Acid treatment of Raw-Maghnite was indicated to cause a relative reduction in the content of octahedrally spaced Al_2O_3 and a relative increase in silica (SiO_2).

Sample	SiO_2	Al_2O_3	Fe_2O_3	CaO	MgO	Na_2O	K_2O	TiO_2	SO_3	PF*
Raw-Mag	69.39	14.67	1.16	0.30	1.07	0.50	0.79	0.16	0.91	11
Mag-H	71.70	14.03	0.71	0.28	0.80	0.21	0.77	0.15	0.34	11

Table 1. Elementary compositions of Protons exchanged samples "Maghnite" (Compositions wt%). PF* : Pert in Fire.

Table 2 shows the various types of montmorillonites studied, and it can be seen that Maghnite has 11.9 % more SiO_2 than Wyoming 19.35 and Montmorillon (Vienne, French) (Kerr et al., 1950; Damour et al., 1987). When treated with sulfuric acid, this difference is even greater; 14.21 % and 21.66 % as compared to Wyoming and Vienne Bentonite, respectively. Maghnite contains 5.60 % and 5.49 % less Al_2O_3, than the Wyoming and Vienne clay, respectively.

Sample	SiO_2	Al_2O_3	Fe_2O_3	FeO	CaO	MgO	Na_2O	K_2O	TiO_2	SO_3
Wyoming (USA)	50.04	20.16	0.68	00	1.46	0.23	Tr	1.27	00	00
Vienne (Frensh)	57.49	20.27	2.92	0.19	0.23	3.13	1.32	0.28	0.12	00
Raw-Mag (Algeria)	69.39	14.67	1.16	00	0.30	1.07	0.5	0.79	0.16	0.91
Mag-H (Algeria)	71.70	14.03	0.71	00	0.28	0.80	0.21	0.77	0.15	0.34

Table 2. Composition in wt% of American, French and Maghnia Algerian montmorillonite.

Fig. 1. shows the X-ray diffraction patterns of Raw-Maghnite and Mag-H^+. The basal spacing of the Raw-Maghnite was 15.02 Å. The titration of Raw-Maghnite with 0.25 H_2SO_4 resulted in the exchange of exchangeable cations for H^+ in the interlayer.

Damour et al. (1987), Kerr et al. (1950) and Kwon et al, (2001) reported that the decrease in the basal spacing indicates a loss of the interlayer water upon the replacement of Na^+ for H^+. In particular, although the X-ray peak of the montmorillonite did not change substantially before or after the acid treatment, there was a decrease in the basal spacing. This implies that the original structure was well preserves after the acid treatment (Table 3).

The effects of the acid activation process on the FTIR spectrum of the treated Maghnite (Fig. 2) are summarised as follows: The intensity of the absorption band at 3630 cm^{-1} (Al-Al-OH coupled by Al-Mg-OH stretching vibrations) decreases with acid treatment. The bands at 3425 cm^{-1} and 3200 cm^{-1} (absorption of interlayer water) become more diffuse with acid treatment (Farmer, 1974). The intensity of the Si-O out of plane and Si-O-Si (2 bands) in plane stretching bands at 1116, 1043 and 999 cm^{-1} have not been affected by acid treatment. The Al-Al-OH (920 cm^{-1}), $AlFe^{3+}OH$ (883 cm^{-1}) and Al-Mg-OH (846 cm^{-1}) deformation bands decrease with acid treatment. The intensity of the band at 796 cm^{-1} increases with treatment, reflects alterations in the amount of amorphous silica in accordance to the findings of others workers (Farmer 1979; Breen et al., 1995). The intensity of the band at 628 cm^{-1} (either Al-OH or Si-O bending and/or Al-O stretching vibration) gradually decreases with acid treatment in good agreement with the findings by Komadel, (2003). The intensity of the band at 467 cm^{-1} (Si-O-Al and Si-O-Mg coupled by OH vibration or Si-O bending vibrations) is essentially unchanged.

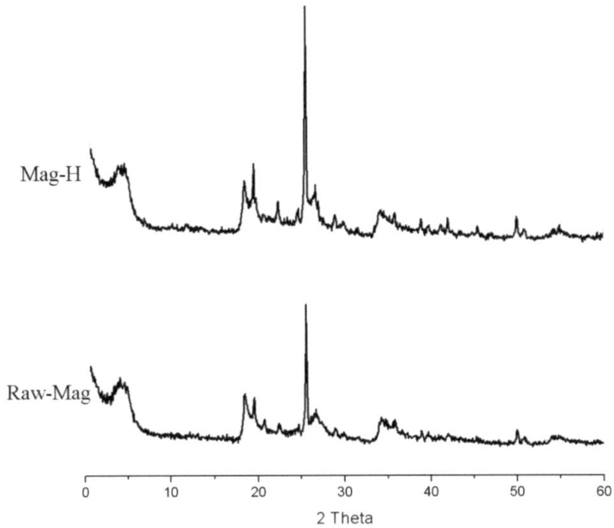

Fig. 1. X-ray Diffraction of Raw-Maghnite and Mag-H.

Samples	dhkl (A°)	Hkl	Nature of sample
Raw-Mag	12.50	001	Montmorillonite
	4.47	110	Montmorillonite
	4.16	//	Quartz
	3.35	//	Quartz
	3.21	//	Feldspath
	3.03	//	Calcite
	2.55	200	Montmorillonite
	1.68	009	Montmorillonite
	1.49	060	Montmorillonite
Mag-H	15.02	001	Montmorillonite
	4.47	110	Montmorillonite
	4.16	//	Quartz
	3.35	//	Quartz
	3.21	//	Feldspath
	3.03	//	Calcite
	2.55	200	Montmorillonite
	1.68	009	Montmorillonite
	1.49	060	Montmorillonite

Table 3. XRD characteristic of Raw-Maghnite and Mag-H.

27Al NMR spectra of both Raw-Maghnite and Mag-H 0.25 M are given in Fig. 3. The spectra of Maghnite exhibits mainly the typical resonance at 2.9 ppm of octahedral aluminium (6Al) in a phyllosilicate but also a small but significant contributions at 60 and 68 ppm assigned to aluminium tetrahedrally co-ordinated to oxygen (4Al) (Samajovà et al., 1992; Benharrats et al., 2003).

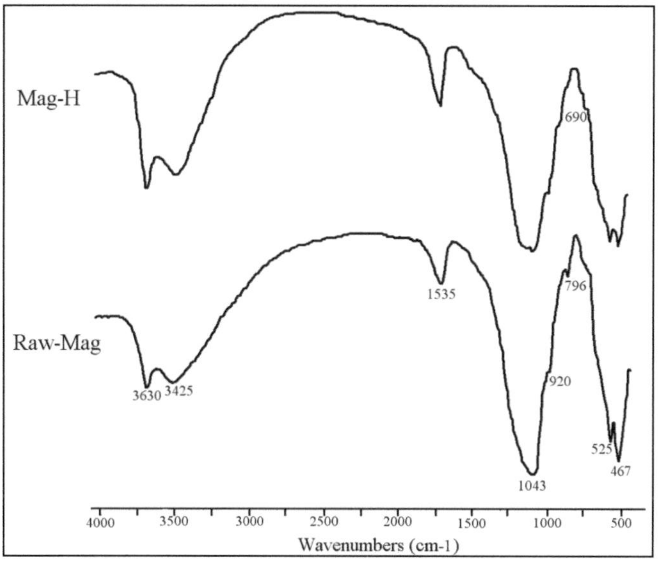

Fig. 2. IR Spectra of untreated Clay (Raw-Maghnite) and Acid treated Clay (Mag-H 0.25M).

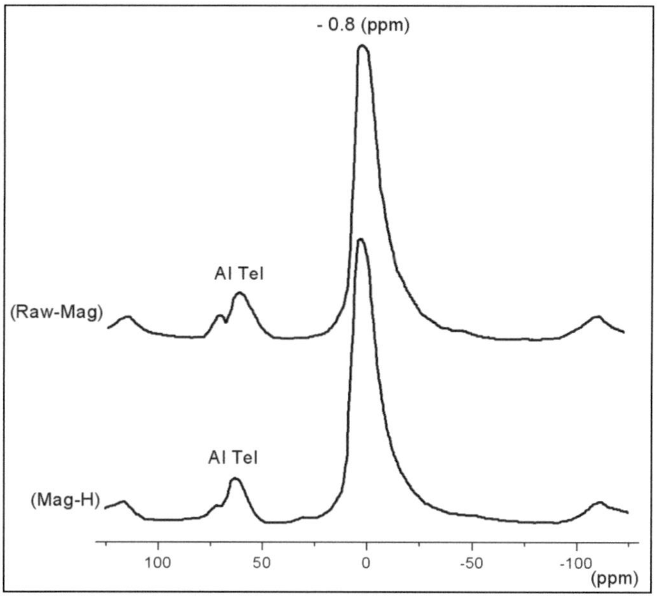

Fig. 3. 27Al MAS NMR spectra of Raw-Maghnite and Mag-H.

The 29Si MAS NMR spectra for the Raw-Maghnite and Mag-H 0.25M are shown in Fig. 4. The dominant resonance at -93.5 ppm corresponds to Q3 (O-Al) units, i.e. SiO4 groups cross linked in the tetrahedral sheets with no aluminium in the neighbouring tetrahedral (Tkàc et al., 1994). The resonance at -112 ppm corresponds to three-dimensional (3D) silica with no aluminium present, designed Q4 (O-Al) (Tkàc et al., 1994; Kwon et al., 2001).

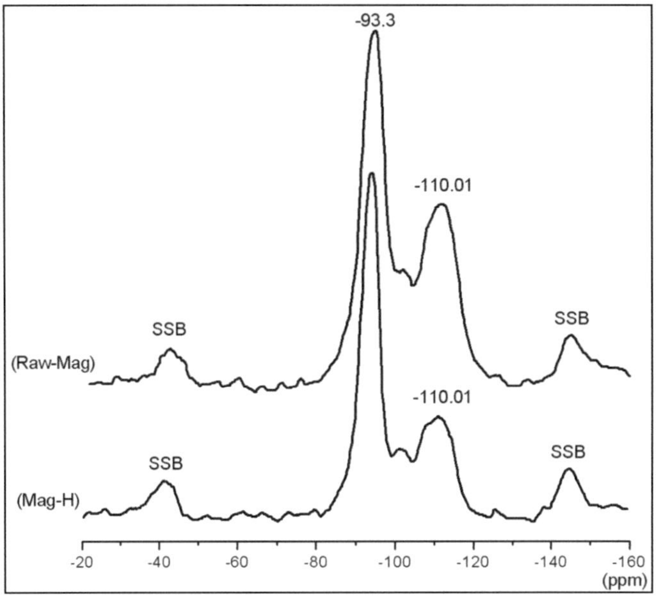

Fig. 4. 29Si MAS NMR spectra of Raw-Maghnite and Mag-H.

2.3 Chemical synthesis

To obtain the polymer, 0.022 moles of monomers (aniline and/or o-toluidine) was added to 1 g of the clay and the mixture was kept under magnetic stirrer at room temperature for 1 hour. The chemical polymerization began when 100 ml of 0.1 M ammonium persulfate solution was slowly added to the mixture (thus, the final concentration of aniline and o-toluidine is 0.022 M). The reaction was carried out under magnetic stirring for 24 hours at room temperature and then the solid product was filtered under vacuum and washed with perchloric acid and distilled water to remove all traces of oxidant and oligomers. All the materials were dried at 60°C. The polymers remain into the catalyst layers forming a polymer at the last of the synthesis. However, the Mag-H is easy to separate by filtration because it is insoluble in the solvents in which the polymers are soluble (Njopwouo et al., 1988).

The same procedure was used for the copolymer synthesis, keeping always the mass ratio Mag-H/monomers (aniline and/or o-toluidine were added with molar ratios: 20/80, 50/50 and 80/20 to a total concentration of 0.22 M.) at 0.5.

2.4 Copolymer characterization

The X-ray diffraction of the powder nanocomposites were taken using a Bruker CCD-Apex equipment with a X-ray generator (Cu Ka and Ni filter) operated at 40 kV and 40 mA. X-ray fluorescence spectroscopy of the powder nanocomposites was made using a Philips PW1480

equipment with a UNIQUANT II software to determine elements in a semi quantitative way.
The XPS spectra were measured with a VG-Microtech Multilab electron spectrometer using non-monochromatised Mg Kα (1253.6 eV) radiation from a twin anode source operated at 300W (20 mA, 15 kV). Photoelectrons were collected into a hemispherical analyser working in the constant energy mode at pass energy of 50 eV. The binding energy (BE) of the Cls peak at 286.4 eV was taken as internal standard. Peak analysis was done with mixed Gaussian/Lorentzian function lineshape by using the Peak-fit program implemented in the control software of the spectrometer. The pressure in the analysis chamber was always lower than 2.10^{-9} Torr. Thermal degradation of the samples does not occur because the time required to collect the XPS spectra is low.
The XPS measurements have been performed in the Mag-H/polymer samples. That is, the polymer has not been separated from the catalyst.
For recording the UV-Vis absorption spectra, a Hitachi U-3000 spectrophotometer was used. The solution of the copolymer in N-methyl-2-pyrrolidone (NMP) was used for recording the spectrum. Fourier transform infrared (FT-IR) spectroscopy was recorded using a Bruker Alpha.
For TEM experiments, the polymers were dispersed in water and then cast in TEM grids. The images were collected using a JEOL (JEM-2010) microscope, working at an operation voltage of 200 kV.

2.5 Conductivity

Conductivity measurements were carried out using a Lucas Lab resistivity equipment with four probes in-line. The samples were dried in vacuum during 24 h; and pellets of 13mm diameter were prepared using a FTIR mold by applying a pressure of 10 Tn/cm^2.

2.6 Electrochemistry

The electrochemical behaviour of the polymers was studied by cyclic voltammetry after their extraction from the composite by dissolving in the N-methyl-2-pyrrolidone (NMP). It is known that this kind of conducting polymers is soluble in NMP (Yoshimoto et al., 2004; Sung et al., 2005), while the clay remains in solid state. Thus, both components can be separated by filtration. The electrochemical measurements were carried out using a conventional cell of three electrodes. The counter and reference electrodes were a platinum foil and a hydrogen reversible electrode (RHE) immersed in the same electrolyte, respectively. The working electrode was prepared as follows: after the polymer was extracted from the polymer using NMP, 50 μL of this solution were cast over graphite carbon electrodes and the solvent evaporated to create polymeric films. The electrolyte used was 1M $HClO_4$ and all experiments were carried out at 50 mV s^{-1}.

3. Results and discussion

3.1 Copolymer characterization

3.1.1 XRD

The Raw-Maghnite (Raw-Mag), the protonated Maghnite (Mag-H) and PoT/Mag-H nanocomposites were characterized using X-ray diffraction to check changes in the interlayer spacing (Fig. 5). The XRD patterns show that the (001) diffraction peak between 3.5° and 5.5° changes depending on the protonated and the polymer intercalated.

Table 4. includes the d-spacing between the montmorillonite sheets calculated from the Bragg equation, the basal spacing and the maximum 2θ of the peaks; that expressed by law :

$$n\lambda = 2d\sin\theta$$

where λ is the X-ray wavelength, d is the spacing between the atomic planes, and θ the angle between the X-ray beam and the atomic plane. Constructive interference occurs for integer values of n. By measuring θ for a known wavelength the Bragg spacing d can be determined.

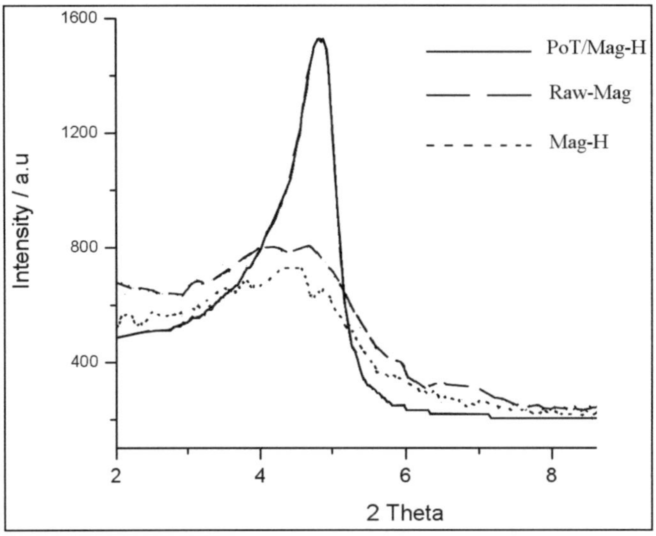

Fig. 5. XRD diffraction patterns of Raw-Maghnite and Protonated Maghnite (Mag-H) and PoT/Mag-H nanocomposite.

Polymers	Peak maximum, 2θ max (deg)	Basal spacing, $d_{(001)}$ (Å)	Interlayer spacing, Δd (Å)
Raw-Mag	4.5	13.4	3.5
Mag-H	4.4	13.2	3.3
PoT/Mag-H	4.9	14.1	4,6
P(Ani-*co*-oT)/Mag-H (20-80 [a])	4.8	13.9	4.5
P(Ani-*co*-oT)/Mag-H (50-50 [a])	4.9	13.9	4.5
P(Ani-*co*-oT)/Mag-H (80-20 [a])	4.8	14.1	4.6

Table 4. Peak maximum and d-spacing of protonated and the nanocomposites intercalated into Raw-Maghnite. [a]Feed composition corresponding to aniline/o-toluidine ratio in the reactor.

3.1.2 XPS

Poly(o-toluidine), polyaniline and copolymer of them at different feed ratio (20/80, 50/50 and 80/20 aniline/o-toluidine ratio) were obtained using the procedure described in the experimental section and characterized by X-ray photoelectronic spectroscopy (XPS).

Fig. 6. shows the XPS spectra for poly(o-toluidine) synthesized using both the conventional and our method. The C1s signal for poly(o-toluidine) prepared using Mag-H (PoT/Mag-H) (Fig. 6a.) is practically identical to those obtained in perochloric acid (PoT/$HClO_4$) (Fig. 6b.). The C1s of poly(o-toluidine) can be deconvoluted into two peaks, the first one at 284.1 eV is assigned to aromatic carbons. The peak at 285.8 eV could be attributed to carbon atoms in quinonimine units and charged species (Chan et al., 1993; Belbachir et al., 2001) such as -C-N^+ and -C=N^+. However, this peak could correspond also to carbon atoms on the methyl group.

The N1s signal for both PoT/Mag-H (Fig. 6c.) and PoT/$HClO_4$ (Fig. 6d.) can be deconvoluted into two principal peaks attributed at both neutral amino or imine groups (399.2 eV) or positively-charged nitrogen atoms (402 eV). In Table 5, the ratio between charged and neutral nitrogen atoms (N^+/N) increases remarkably when a small quantity of aniline is added to the feed. In polyaniline substituted with anionic groups, the N^+/N ratio is related to either the doping level or the charge localization (Chan et al., 1993; Salavagione et al., 2004) however, the increase of N^+/N in this case is only related to the increase in the doping level because the substituting group (-CH_3) is not an ionic species and therefore it is unable to fix charge.

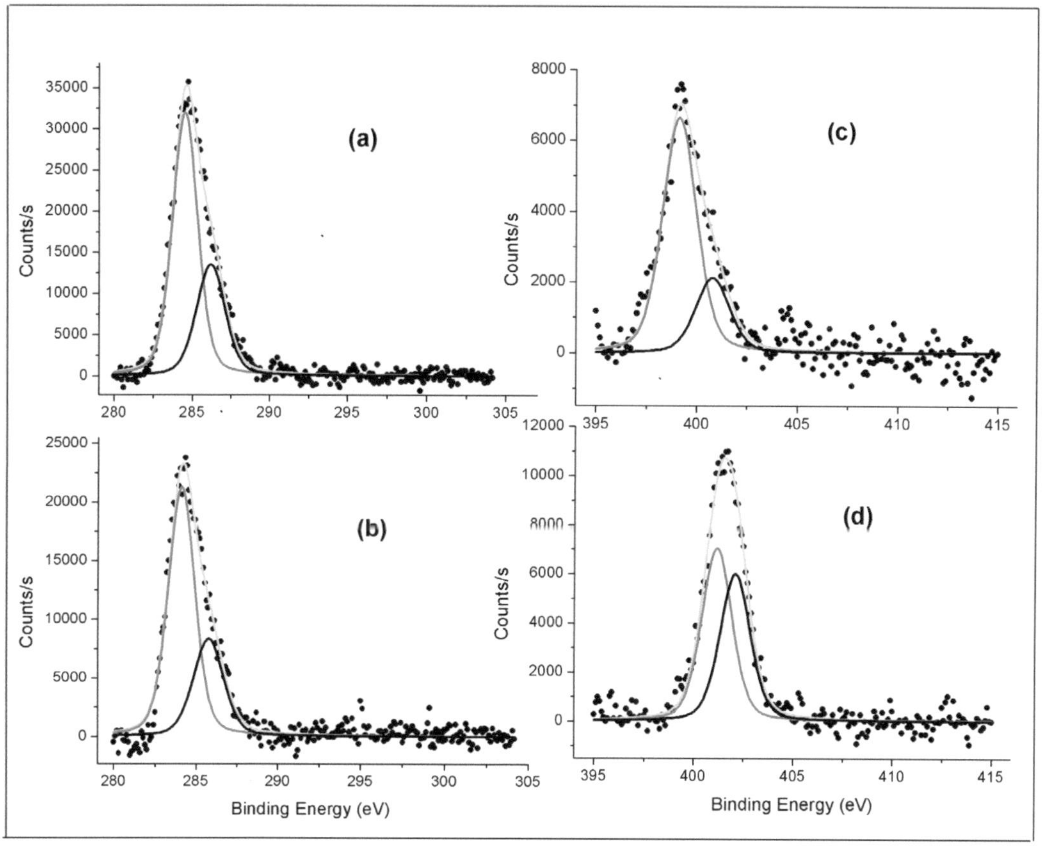

Fig. 6. XPS of poly(o-toluidine) for both C1s of PoT/Mag-H (a) and PoT/$HClO_4$ (b) and for N1s signal of PoT/Mag-H (c) and PoT/$HClO_4$ (d).

Table 5. shows the relative areas of de C1s peak for PoT and copolymers of aniline and o-toluidine at different feed ratio. Here, it can be noted that while the area of the peak at 284.1 eV decreases when the amount of o-toluidine in the feed increase (column C1/C in Table 5); however, the area of the peak at 285.8 eV increases (column C2/C in Table 5). Therefore, this signal at 285.8 eV has contribution of carbon linked to the methyl group and the peak at 284.1 eV only corresponds to carbon atoms in quinoneimine units.

Polymers	C1s Signal areas		N1s Signal areas
	C_1/C	C_2/C	N^+/N
P(Ani-*co*-oT)/Mag-H (20-80 [a])	0.626	0.374	0.486
P(Ani-*co*-oT)/Mag-H (50-50 [a])	0.656	0.343	0.511
P(Ani-*co*-oT)/Mag-H (80-20 [a])	0.684	0.315	0.666
PoT/MagH	0.673	0.327	0.320
PoT/$HClO_4$	0.689	0.311	0.397

Table 5. Areas of the XPS signal of C1s and N1s for homopolymers of o-toluidine and copolymer of them. C refers to total carbons area, C1 refers to BE 284.1 eV and C2 correspond to 285.8 eV. [a]Feed composition corresponding to aniline/o-toluidine ratio in the reactor.

3.1.3 FT-IR spectroscopy

The FTIR spectra of homopolymers and copolymers are shown in Fig. 7. The spectral data of samples are in Table 6.

The FT-IR spectrum of PoT/Mag-H exhibits the following main spectral features (Kim et al., 1988; Gruger et al., 1994; Andrade et al., 1996) :

i. The broadband at 1580 cm^{-1} can be assigned to the C-C stretching mode combined with C=N stretching vibrations of the quinoid rings.
ii. The band at 1496 cm^{-1} is attributed to the C-C stretching modes in the benzenoid units.
iii. The bands at 1325 cm^{-1} and 1223 cm^{-1} are assigned to the C-N stretching vibrations in the polymer chain.
iv. The band at 1158 cm^{-1} is related to the C=N stretching mode; this band is a characteristic of doped PoT. It can be due to the doping of the PoT by the ambient gases.
v. The band at 1109 cm^{-1} can be attributed to charge delocalization on the polymer backbone.
vi. The band located at 804 cm^{-1} represents the paradisubstituted benzenoid rings in PoT.

In addition, the presence of quinoid and benzenoid bands (1572 and 1515 cm^{-1}) shows that the PoT is composed of imine and amine units. As can be seen, the spectra for both polymers are almost identical and show most of the bands of very similar conducting polymers (Sariciftci et al., 1990).

Wavenumber (cm^{-1})		Band characteristics
PoT/$HClO_4$	PoT/Mag-H	
579	582	C–H out of plane bending vibration
811	807	Paradisubstituted aromatic rings indicating polymer formation
878	871	Due to the methyl group attached to the phenyl ring
1109	1092	C–H in plane bending vibration
1160	1157	Vibration band of the dopant anion
1315	1325	Aromatic C–N stretching indicating secondary aromatic amine group
1496	1507	C–N stretching of benzenoid rings
1582	1578	C–N stretching of quinoid rings

Table 6. Characteristic frequencies of chemically synthesized poly(o-toluidine) synthitezed with $HClO_4$ and with Mag-H.

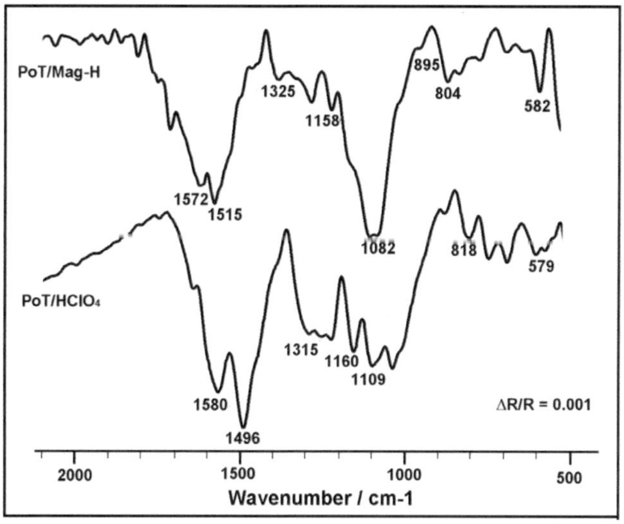

Fig. 7. FT-IR adsorption spectra of the of poly(o-toluidine) doped with Mag-H and with $HClO_4$.

The spectra of copolymer is shown in Fig. 8, they present features common to homopolymers synthesized using Mag-H. In the spectra of copolymer appears absorption band at 1109 cm^{-1} (C-H bending) and 1496 cm^{-1} (C-C stretching) indicating the existence of methyl group in the copolymer; besides this, new absorption band appears at 1109 cm^{-1}, which increases the intensity of the copolymer with increase of the o-toluidine content in the comonomer feed.

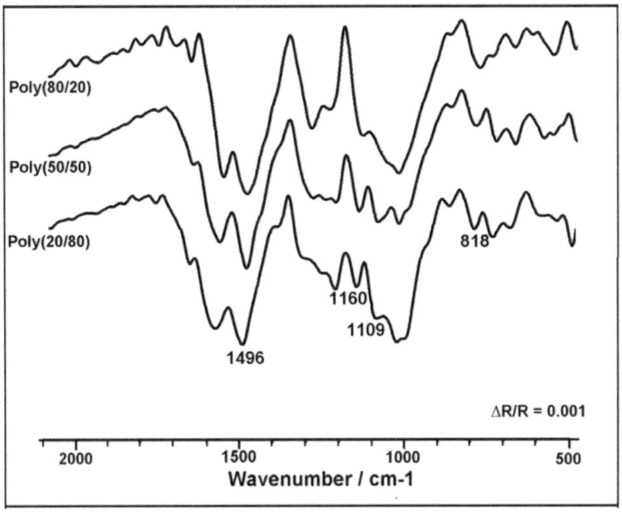

Fig. 8. FT-IR adsorption spectra of the copolymers with (aniline/o-toluidine) molar ratios (20/80, 50/50 and 80/20).

3.1.4 UV–Vis absorption measurement for the homo- and co-polymers

The absorption maxima for the PoT/Mag-H, PoT/$HClO_4$ and poly(ANI-co-oT)/Mag-H in DMSO solution are given in Table 7, it is observed that, there are two absorption bands in the electronic spectra of homopolymers and copolymers (Fig. 9). The band around 310 nm is assigned to π–π* which corresponds to band gap and the band above 600 nm is assigned to n–π* exciton band or inters band charge transfer associated with the excitation of benzenoid to quinoid rings (Chan et al., 1993; Yang et al., 2010).

Polymers	Bandgap *Adsorption bands λ_{max} (nm)*	Exciton *Adsorption bands λ_{max} (nm)*
P(Ani-*co*-oT)/Mag-H (20-80 [a])	311	596
P(Ani-*co*-oT)/Mag-H (50-50 [a])	316,5	599,5
P(Ani-*co*-oT)/Mag-H (80-20 [a])	307,5	597
PoT/MagH	311,5	600,5
PoT/$HClO_4$	309	600

Table 7. UV-Vis. absorption bands of polymers synthesized using Mag-H and $HClO_4$. [a]Feed composition corresponding to (aniline/o-toluidine) ratio in the reactor.

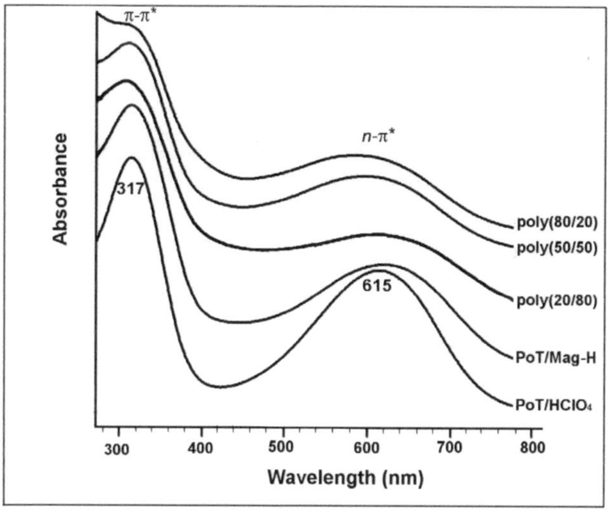

Fig. 9. UV–vis spectra of the poly(o-toluidine)/$HClO_4$ and poly(o-toluidine)/Mag-H.

The π–π* transistion of copolymers shows hyposochromic shift (blue shift). This may be due to the increase the degree of freedom and entropy of solvation. The n–π* exciton band shows a bathochromic shift (red shift) with an increase in dielectric constant of the solvent (Kim et al., 1988). The π–π* band in copolymer shifts to the lower wavelength (higher energy) as the mole percent of o-toluidine in copolymer increases, which may be due to addition of more o-toluidine unit, which twist the torsion angle, which is expected to increase the average bandgap in conjugated polymer chain.

3.2 Copolymer properties

3.2.1 Conductivity

The conductivities of homopolymers and copolymers are given in Table 8. The values lie in between 2.19 x 10^{-5} and 2.24 x 10^{-3} $S.cm^{-1}$. The conductivity of chemically synthesized PoT/$HClO_4$ salt show very low conductivity (2.02 x 10^{-3} $S.cm^{-1}$) is higher than that of PoT/Mag-H salt (2.19 x 10^{-5} $S.cm^{-1}$).

Polymers	Conductivity ($S.cm^{-1}$)
P(Ani-*co*-oT)/Mag-H (20-80 [a])	2.24 x 10^{-3}
P(Ani-*co*-oT)/Mag-H (50-50 [a])	1.52 x 10^{-4}
P(Ani-*co*-oT)/Mag-H (80-20 [a])	6.26 x 10^{-5}
PoT/MagH	2.19 x 10^{-5}
PoT/$HClO_4$	2.08 x 10^{-3}

Table 8. The conductivity values of polymers synthesized with $HClO_4$ and Mag-H. [a]Feed composition corresponding to (aniline/o-toluidine) ratio in the reactor.

The conductivity of all the copolymers was found to be higher than that of the homopolymers PoT/Mag-H (2.19 x 10^{-5} $S.cm^{-1}$). The conductivity increases with aniline content. The higher conductivity of the copolymers compared to that of the PoT/Mag-H homopolymers indicates lowering of the band gap in the copolymers formed with toluidine (with a donor $-CH_3$ group). This finding is supported by the bathochromic shift of the peak around 600 nm in the electronic absorption spectra of the copolymers as discussed later. However, the steric effect of the methyl group is likely to disrupt the overlapping of orbitals, hence lowering the degree of conjugation, by forcing the aromatic rings out of plane relative to each other. The poorer polaron formation of the copolymers with the higher concentration of o-toluidine is consistent with the conductivity decrease of copolymers. These results are reasonable because the self-doping degree must decrease as the aniline mole fraction increases. Therefore, we can conclude that all copolymers have some degree of self-doping despite the lower reactivity of o-touluidine monomer.

Also it is also possible that the addition of aniline in the ratio of copolymers curing affects the chain alignment of the polymer, which leads to the increase of conjugation length and that brings about the increase of conductivity.

3.2.2 Solubility

The solubility of PoT/Mag-H, PoT/$HClO_4$ and these copolymers of aniline/o-toluidine in different solvents are show in Table 9. The solubility decreases with increasing the concentration of aniline units in the copolymer. The solubility improvement of polymers prepared with Mag-H compared with $HClO_4$ must be related to the presence of anionic and/or cationic groups (from the natural sodium montmorillonite clay units) in the synthesis of polymers.

Samples	NH_4OH	DMSO	THF	NMP	Etha-nol	CH_3COOH	Tolue-ne	Acet-one	CCl_4	C_6H_6
P(Ani-*co*-oT)/Mag-H (20-80[a])	S/BV	PS/BR	PS/BG	PS/BR	IS	PS/B	PS/BG	IS/BR	IS	IS
P(Ani-*co*-oT)/Mag-H (50-50[a])	S/BV	S/BV	PS/BR	S/BR	PS/BV	PS/BG	IS	PS/BR	IS	IS
P(Ani-*co*-oT)/Mag-H (80-20[a])	S/V	S/BV	S/BR	S/BV	PS/B	S/V	IS	PS/BR	PS/B	IS
PoT/MagH	S/V	S/BV	S/BV	S/BV	PS/BV	S/V	IS	PS/BR	PS/B	IS
PoT/$HClO_4$	S/V	S/BR	S/BR	S/BR	PS/RC	PS/B	IS	PS/V	IS	IS

Table 9. Solubility of PoT/Mag-H, PoT/$HClO_4$ and copolymers (Aniline/o-Toluidine) in different solvents. Solubility in solvents (IS = Insoluble ; PS = Partially ; S : Soluble). The solution color is indicated in the parentheses with the following abbreviations (V : Violet ; BV : Bluish violet ; BR : Brownish red ; BG : Bluish green ; B : Blue).

The solubility of the resulting conducting polymer of (Aniline/o-Toluidine) increases with increasing the feed ratio of o-toluidine in the polymerization, and it was observed that the highest solubility gives a low conductivity. That is to say when the solubility decreases the conductivity increases. This result indicates the solubility of polymers synthesized with Mag-H decreased with increasing polymer chain.

3.2.3 Electrochemical response

To study the electrochemical response of polymers and copolymers, a conventional cell of three electrodes was used. To separate the polymers from the catalysts, they were dissolved in N-methyl-2-pyrrolidone (NMP). Films of the polymers were made by casting a drop of a solution of them in NMP over the electrode and heating with an infrared lamp to remove the solvent.

Fig. 10. shows the voltammograms obtained with the two electroactive poy(o-toluidine) films in 1M $HClO_4$ solution. The current densities have been normalized in order to clearly compare between the two polymers, due to the thickness of the films could vary from different samples. The electrochemical behaviour of the PoT/Mag-H exhibits two anodic peak around 0.48V and 0.71V however, on the reverse scan two reduction peaks around 0.59 V and 0.40 V are observed. This behaviour differs from those obtained when the polymer is created in acidic conditions (PoT/$HClO_4$, Fig. 10. continuous line). The polymer created in $HClO_4$ shows the similar electrochemical response reported by Borole et al. (2003, 2004, 2006), two pairs of redox peaks were detected with a potential shift at 0.58/0.37V and 0.73/0.65V.

To try to clarify all peaks observed, we have synthesized polyaniline and copolymer of aniline/o-toluidine with Mag-H as proton source. The cyclic voltammogram of these copolymers obtained by Mag-H is shown in Fig. 11. The voltammetric behaviour of copolymer (aniline/o-toluidine) (20/80) is similar to those obtained in PoT/Mag-H. At higher concentration of aniline monomers (50/50), two pairs of anodic and cathodic current peaks are seen clearly, centred at around 0.51 and 0.80V. At lower electrode potentials, not exceeding 0.51V, the reduced (leucoemeraldine) form prevails, whereas at high potentials, exceeding ca. 0.80V, the fully oxidized (pernigraniline) form exists, leaving a broad potential window, ranging from 0.51 to 0.80 V, where the conducting emeraldine form is most stable. In addition, for copolymers with high amounts of aniline (80/20) shows the same behaviour (tow redox process). Therefore, besides that the properties of this kind of conducting polymers do not depend on the type of anion used in the synthesis (Chan 1993), the electrochemical properties depend strongly on the proton source (inorganic acids or Mag-H catalyst).

3.2.4 TEM

Fig. 12. show the transmission electron micrographs of two typical polymers PoT/Mag-H and PoT/$HClO_4$. Transmission electron micrographs of the PoT/Mag-H reveal a structure similar to those of PoT/$HClO_4$. Presence of chopped fibers in the two polymers can be explained by considering the steric contribution of the methyl group. The methyl group present at the ortho position of the benzene ring leads to a distortion in the polymer chains and restrict the polymer growth in a linear fashion, which in turn results in a breakdown of the polymer chain into small fragments and this appears as small chopped or needle shaped fibers in the micrograph. Interestingly the polymer of PoT/Mag-H is more chopped fibers comprising of a whole polymer matrix than the corresponding PoT/$HClO_4$. The interplanar distances, however, do not exhibit any perceptible change with the two different methods of synthesis. The PoT/Mag-H apparently has larger interplanar distances than the corresponding PoT/$HClO_4$.

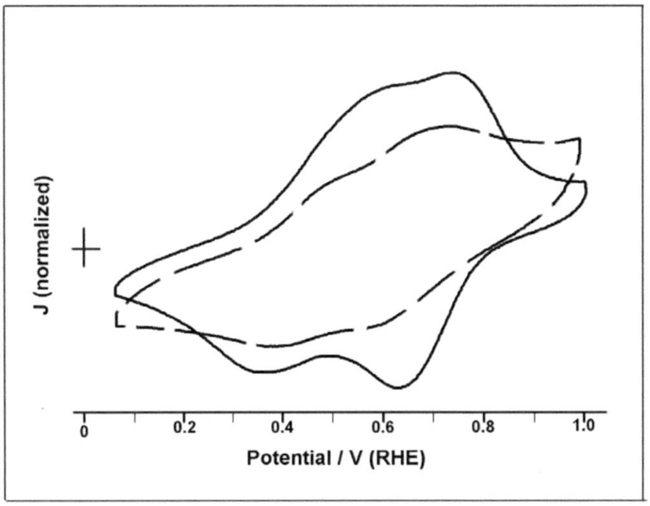

Fig. 10. Cyclic voltammograms recorded for a graphite electrode covered by PoT/Mag-H (dashed line) and PoT/$HClO_4$ (solid line) in 1M $HClO_4$ solution. Scan rate 50 mV s^{-1}.

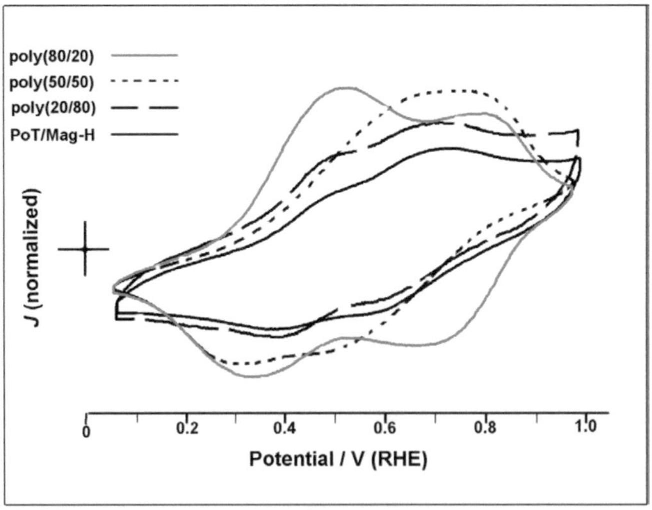

Fig. 11. Cyclic voltammograms recorded for a graphite carbon electrode covered by copolymers the feed ratio (20/80, 50/50 and 80/20 aniline/o-toluidine ratio) in 1M $HClO_4$ solution. Scan rate 50 mV s^{-1}.

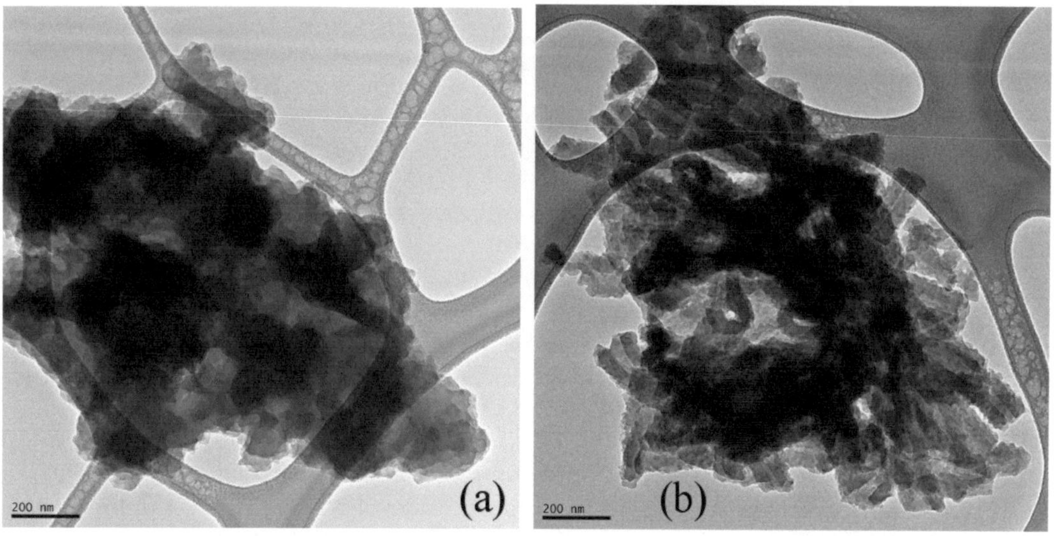

Fig. 12. Transmission electron micrographs (TEM) of poly(o-toluidine) doped with $HClO_4$ (PoT/$HClO_4$) (a) and doped with Mag-H (PoT/Mag-H) (b).

4. Conclusions

We have developed a new synthetic method to produce conducting polymers, which does not use inorganic acid but a non-toxic clay catalyst as proton source. Preliminary results about the spectroelectrochemical properties have been shown. However, we can conclude that the electrochemical behaviour of PoT/Mag-H is quite different from those observed for PoT/$HClO_4$. The polymers synthesized using Mag-H could show a structure that is a mixture of conducting (polyaniline-type) and redox (phenazine-type) units.
The degree of d-spacing of PoT/Mag-H nanocomposites was expanded to 4.6 (Å) compared to the Mag-H 3,3 (Å), this result shows the strong interaction between PoT and Mag-H that gives additional stability to the intercalated polymer within the matrix. The nanocomposite is having good conductivity, but less than that of pure PoT, due to the weakening in interchain interactions within the confined matrix. In addition, the PoT/Mag-H is more soluble in different solvents than PoT/$HClO_4$.

5. Acknowledgements

This work was supported by The National Agency for the Development of University Research (ANDRU), the Directorate General of Scientific Research and Technological Development (DGRSDT) of Algeria and the Departamento de Química Física e Instituto Universitario de Materiales, Universidad de Alicante (Spain).

6. References

Andrade, E. M. Molina, F. V. Florit, M. I. & Posadas, D. (1996), Ir study of the first redox couple of poly(*o*-toluidine). anion involvement and potential dependence in strongly acidic solutions, *Journal of Electroanalytical Chemistry*, Vol.415, No.1-2, (October 1996), pp. 153-169, ISSN 1572-6657.

Barbero, C. Salavagione, H.J. Acevedo, D.F. Grumelli, D.E. Garay, F. Planes, G.A. Morales, G.M. & Miras, M.C. (2004),Novel synthetic methods to produce functionalized conducting polymers I. Polyanilines, *Electrochimica Acta,* Vol.49, No.22-23, (September 2004), pp. 3671-3686, ISSN 0013-4686.

Barrios, E. M. Mulica, G. A. & Velasquez, C. L. (2006), Studies of the presence of dicarboxylic acids in the electrochemical synthesis of poly(aniline): Case poly(itaconic acid), *Journal of Electroanalytical Chemistry,* Vol.586, No.1, (January 2006), pp. 128-135, ISSN 1572-6657.

Belbachir, M. & Bensaoula, A. (2001), Composition and Method for Catalysis using Bentonites, U.S. Patent, No. 7094823, (March 2002), ISBN 9780199751105, USA.

Benharrats, N. Belbachir, M. Legran, A. P. & Déspinose de le Caillerie, J. B. (2003), 29Si and 27Al MAS NMR study of the zeolitization of kaolin by alkali leaching, *Clay Minerals,* Vol.38, No.1, (March 2003), pp.49-61, ISSN 0009-8558.

Benyoucef, A. Boussalem, S. Ferrahi, M.I. & Belbachir, M. (2010), Electrochemical polymerizationa nd *in situ* FTIRS study of conducting polymers obtained from *o*-aminobenzoic with aniline at platinum electrodes, *Synthetic Metals,* Vol.160, No.15-16, (August 2010), pp. 1591-1597, ISSN 0379-6779.

Bergeron, J.Y. & Dao, L.H. (1991), Poly(aniline-co-N-butylaniline) copolymers. A new internal electrically conducting composite, *Polym. Commun.* Vol.32, No.13, (May 1991), pp. 403-404, ISSN 0263-6476.

Blomquist, M. Lindfors, T. Latonen, R.M & Bobacka, J. (2009), Electropolymerization of *N*-methylanthranilic acid and spectroelectrochemical characterization of the formed film, *Synthetic Metals,* Vol.159, No.1-2, (January 2009), pp. 96-102, ISSN 0379-6779.

Borole, D.D. Kapadi, U.R. Mahulikar, P.P. & Hundiwale, D.G. (2003), Studies on electrochemical, optical and electrical conductivity characteristics of copolymer of polyaniline-co-poly(*o*-toluidine) using various organic salts, *Materials Letters,* Vol.57, No.22-23, (July 2003), pp. 3629-3635, ISSN 0167-577X.

Borole, D.D. Kapadi, U.R. Mahulikar, P.P. & Hundiwale, D.G. (2004), Electrochemical behaviour of polyaniline, poly(o-toluidine) and their copolymer in organic sulphonic acids, *Materials Letters,* Vol.58, No.29, (November 2004), pp. 3816-3822, ISSN 0167-577X.

Borole, D.D. Kapadi, U.R. Mahulikar, P.P. & Hundiwale, D.G. (2006), Electrochemical synthesis and characterization of conducting copolymer: Poly(o-aniline-co-o-toluidine), *Materials Letters,* Vol.60, No.20, (Septembre 2006), pp. 2447-2452, ISSN 0167-577X.

Breen, C. Madejová, J. & Komadel, P. (1995), Characterisation of moderately acid-treated, size-fractionated montmorillonites using IR and MAS NMR spectroscopy and thermal analysis, *J. Materials Chemistry,* Vol.5, No.3, (dec 1995), pp. 469-474. ISSN 1364-5501.

Cha, S.K.J. (1997), Electropolymerization rates of polythiophene/polypyrrole composite polymer with some dopant ions, *Journal of Polymer Science Part B: Polymer Physics,* Vol.35, No.1 (January 1997), pp. 165-172, ISSN 1099-0488

Chan, H.S.O. Ng, S.C. Sim, W.S. Tan, K.L. & Tan, B.T.G. (1993), Synthesis and characterization of conducting poly(o-aminobenzyl alcohol) and its copolymers with aniline, *Macromolecules,* Vol.26, No.1, (January 1993), pp. 144-150, ISSN 0024-9297.

Damour, A. & Salvetat, D. (1987), Analyses sur un hydrosilicate d'alumine trouvé à Montmorillon. *Annales de Chimie et Physique*, Vol.3, No.*21*, (September 1993), pp. *376*–383, ISSN 03651444.

Dhawan, S.K. & Trivedi, D.C. (1993), Influence of polymerization conditions on the properties of poly(2-methylaniline) and its copolymer with aniline, *Synthetic Metals*, Vol.60, No.1, (September 1993), pp. 63-66, ISSN 0379-6779.

Diaz, A.F. & Logan, J.A. (1980), Electroactive polyaniline films, *Journal of Electroanalytical Chemistry*, Vol.111, No.1, (April 1980), pp.111-114, ISSN 0022-0728.

Farmer, V.C. (1974), The Infrared Spectra of Minerals, *Mineralogical Society*, Farmer, V.C. (Ed), 427-485, Mineralogical Society Monograph 4, ISBN 13-978-0903056052. London, UK.

Farmer, V.C. (1979), Infrared spectroscopy, *Pergamon Press*, Van Olphen H. Fripiat, J.J. (Ed), 285–337, Data Handbook for Clay Materials and other Non-metallic Minerals, ISBN 0-56576-010-7, Oxford, UK.

Gningue, D. Horowitz, G. & Garnier, F. J. (1988), Protection of CdSe Oxygen Photoanodes by Poly(bithiophene)-Polypyrrole Composites and Copolymers, *J. Electrochem. Soc*, Vol.135, No. 7, (July 1988), pp. 1695-1699, ISSN 1945-7111.

Gruger, A. Novak, A. Régis, A. & Colomban, P. (1994), Infrared and Raman study of polyaniline Part II: Influence of ortho substituents on hydrogen bonding and UV/Vis-near-IR electron charge transfer, *Journal of Molecular Structure*, Vol.328, No.1, (December 1994), pp. 153-167, ISSN 0022-2860.

Karyakin, A.A. Strakhova, A.K. & Yatsimirsky, A.K. (1994), Self-doped polyanilines electrochemically active in neutral and basic aqueous solutions.: Electropolymerization of substituted anilines, ,*Journal of Electroanalytical Chemistry* Vol.371, No.1-2, (June 1994), pp. 259-265, ISSN 1572-6657.

Kerr, P.F. Hamilton, P.K. & Pill, R.J. (1950), Analytical data on reference, Clay Minerals: American Petroleum Institute Project 49, *Clay Minerals Standards*, preliminary report, Vol.7, No.160 (March 1950), pp. 92-98. ISSN 1572-4352,

Khalkali R.A. (2005), Electrochemical Synthesis and Characterization of Electroactive Conducting Polypyrrole Polymers *Russian Journal of Electrochemistry*, Vol.41, No.9 (September 2004), pp. 1071-1078, ISSN 1572-8838.

Kim, Y.H. Foster, C. Chiang, J. & Heeger, A.J. (1988), Photoinduced localized charged excitations in polyaniline, *Synthetic Metals*. Vol.26, No.1, (October 1988), pp. 49-59, ISSN 0379-6779.

Komadel, P. (2003), Chemically modified smectites Clay Minerals, *Clay Minerals*, Vol.38, No.1, (March 2003), pp.127-138, ISSN 0009-8558.

Kwon, O.Y. Park, K.W. & Jeong, S.Y. (2001), Preparation of Porous Silica-Pillared Montmorillonite: Simultaneous Intercalation of Amine-Tetraethylorthosilicate into H-Montmorillonite and Intra-Gallery Amine-Catalyzed Hydrolysis of Tetraethylorthosilicate, *Bulletin of the Korean Chemical Society*. Vol.22, No.7, (July 2001), pp. 678-686. ISSN 0253-2964.

Laborde, H. Leger, J.-M. Lamy, C. Garnier, F. & Yassar, A. J. (1990), Electrocatalytic oxidation of hydrogen, formic acid and methanol on platinum modified copolymer (pyrrole-dithiophene) electrodes, Journal of Applied Electrochemistry, Vol.20, No.3, (May 1989), pp. 524-526, ISSN 1572-8838.

Langer, J.J. (1993), N-substituted polyanilines: II. Photoacoustic and FT-IR spectra of poly(N-methylaniline) and related copolymers, *Synthetic Metals*, Vol. 35, No.3, (April 1990), pp. 301-305, ISSN 0379-6779.

Li, C. Mu, S. (2005), The electrochemical activity of sulfonic acid ring-substituted polyaniline in the wide pH range, *Synthetic Metals*, Vol.149, No.2-3, (March 2005), pp. 143-149, ISSN 0379-6779.

Lindfors, T. & Harju, L. (2008), Determination of the protonation constants of electrochemically polymerized poly(aniline) and poly(o-methylaniline) films, *Synthetic Metals*, Vol.158, No.8, (April 2008), pp. 233-241, ISSN 0379-6779.

Lindfors, T. Sandberg, H. & Ivaska, A. (2004), The influence of lipophilic additives on the emeraldine base–emeraldine salt transition of polyaniline, *Synthetic Metals*, Vol.142, No.1-3, (April 2004), pp. 231-242, ISSN 0379-6779.

Miller, J. S. (1993), Conducting polymers-materials of commerce. *Advanced Materials*. Vol.5, No.7-8, (July/August 1993), pp. 587–589, ISSN 1521-4095.

Mu, S. & Kan, J. (2002), The electrocatalytic oxidation of ascorbic acid on polyaniline film synthesized in the presence of ferrocenesulfonic acid, *Synthetic Metals*, Vol.132, No.1, (December 2002), pp. 29-33, ISSN 0379-6779.

Mu, S. (2003), The electrocatalytic oxidation of gallic acid on polyaniline film synthesized in the presence of ferrocene phosphonic acid, *Synthetic Metals*, Vol.139, No.2, (December 2003), pp. 287-294, ISSN 0379-6779.

Nalwa, H.S. (1997a), *Handbook of Organic Conductive Molecules and Polymers, Conductive Polymers: Transport, Photophysics and Applications*, ISBN 978-0-471-96275-5 Wiley, New York, USA.

Nalwa, H.S. (1997b), *Handbook of Organic Conductive Molecules and Polymers*, ISBN 978-0-471-96275-5 Wiley, New York, USA.

Njopwouo, D. Roques, G. & Wandji, R. (1988), A contribution to the study of the catalytic action of clays on the polymerization of styrene; II, Reaction mechanism, *Clay Minerals*, Vol.23, No.1, (March 1988), pp.35-43. ISSN 0009-8558.

Novak, P. Muller, K. Santhanam, K.S.V. & Hass, O. (1997a), Electrochemically Active Polymers for Rechargeable Batteries, Chemical Reviews, Vol.97, No.01, (February 05), pp. 207-282, ISSN 1520-6890.

Peters, E.M. & Van Dyke, J.D.J. (1992), Characterization of conducting copolymers of 2,2'-bithiophene and pyrrole by cyclic voltammetry and UV/visible spectroscopy, *Journal of Polymer Science Part A: Polymer Chemistry*, Vol.30, No.9, (August 1992), pp. 1891–1898, ISSN 1099-0518.

Porter, T.L. Caple, G. & Lee, C.Y. (1990), Structural study of the surface of aniline-3-aminophenylboronic acid copolymer films, *Synthetic Metals*, Vol.46, No.1, (January 1992), pp. 105-112, ISSN 0379-6779.

Salavagione, H.J. Acevedo, D.A. Miras, M.C. Motheo, A.J. & Barbero, C.A. (2004), Comparative study of 2-amino and 3-aminobenzoic acid copolymerization with aniline synthesis and copolymer properties, *J. Polymer Science Part A: Polymer Chemistry*, Vol.42, No.22, (November 2004), pp. 5587-5599, ISSN 1099-0518.

Samajovà, E. Kraus, I. & Lajcàkovà, A. (1992), Diagenetic alteration of miocene acidic vitric tuffs of the jastraba formation, *Geologica Carpathica Clays*. Vol.1, No.1, (Mai 1992), pp.21-26, ISSN 1210-2695

San, B. & Talu, M. (1998), Electrochemical copolymerization of pyrrole and aniline, *Synthetic Metals*, Vol.94, No.2, (April 1998), pp. 221-227, ISSN 0379-6779.

Sanchez De Pinto, M.I. Mishima, H.T. & López De Mishima, B.A.J. (1997), Polymers and copolymers of pyrrole and thiophene as electrodes in lithium cells, *Journal of Applied Electrochistry*, Vol.27, No.7, (May 1996), pp. 831-838, ISSN 1572-8838.

Sariciftci, N.S. Kuzmany, H. Neugebauer, H. & Neckel, A. (1990), Structural and electronic transitions in polyaniline: A Fourier transform infrared spectroscopic study, *Journal of Chemical Physics*, Vol.92, No.7, (March 1989), pp. 4530-4540, ISSN 0021-9606.

Sato, M. Yamanaka, S. Nakaya, J. & Hyodo, K. (1994), Electrochemical copolymerization of aniline with o-aminobenzonitrile, *Electrochimica Acta*, Vol.39, No.14, (October 1994), pp. 2159-2167, ISSN 0013-4686.

Skotheim, T.A. Elsenbaumer, R.L. & Reynolds, J.R. (1998), *Handbook of Conducting Polymers*, ISBN 0824757424, Marcel Dekker, New York, USA.

Sun, J.J. Zhou, D.M. Fang, H.Q. & Chen, H.Y. (1998), The electrochemical copolymerization of 3,4-dihydroxybenzoic acid and aniline at microdisk gold electrode and its amperometric determination for ascorbic acid, *Talanta*, Vol.45, No.5, (March 1998), pp. 851-856, ISSN 0039-9140.

Sung, J.H. & Choi, H.J. (2005), Effect of pH on Physical Characteristics of Conducting Poly(o-Ethoxyaniline) Nanocomposites, *Journal of Macromolecular Science, Part B*, Vol.44, No.3, (Feb 2007), pp. 365-375, ISSN 1520-5738.

Talu, M. Kabasakaloglu, M. & Oskoui, R.H.J. (1996), Electrochemical copolymerization of thiophene and aniline, *Journal of Polymer Science Part A: Polymer Chemistry*, Vol.34, No.14, (October 1996), pp. 2981-2989, ISSN 1099-0518.

Tkàc, I. Komadel, P. & Müeller, D. (1994), Acid-treated montmorillonites; a study by 29Si and 27Al MAS NMR, *Clay Minerals*, Vol.29. No.1, (March 1994), pp.11-19, ISSN 0009-8558.

Wei, Y. Focke, W.W. Wnek, G.E. Ray, A. & MacDiarmid, A.G. (1989), Synthesis and electrochemistry of alkyl ring-substituted polyanilines, *Journal of Physical Chemistry*, Vol.93, No.1, (January 1989), pp. 495–499, ISSN 1932-7447.

Wei, Y. Hariharan, R. & Patel, S.A. (1990), Chemical and electrochemical copolymerization of aniline with alkyl ring-substituted anilines, *Macromolecles*, Vol.23, No.3, (February 1990), pp. 758–764, ISSN 1520-5835.

Xu, J.J. Zhou, D.M. Chen, H.Y. & Fang, H.Q. (1997), Amperometric determination of ascorbic acid at a novel 'self-doped' polyaniline modified microelectrode, *Journal of Analytical Chemistry*, Vol.362, No.2, (February 1998), pp. 234-238, ISSN 1608-3199.

Yang, O. Zhang, Y. Li, H. Zhang, Y. Liu, M. Luo, J. Tan, L. Tang, H. & Yao, S. (2010), Electrochemical copolymerization study of *o*-toluidine and *o*-aminophenol by the simultaneous EQCM and in situ FTIR spectroelectrochemisty, *Talanta*, Vol.81, No.1-2, (April 2010), pp. 664-672, ISSN 0039-9140.

Yildiz, H.B. Kiralp, S. Toppare, L. Yagci, Y. & Ito, K. (2006), Synthesis of conducting copolymers of thiophene capped poly(ethylene oxide) with pyrrole and thiophene, *Materials Chemistry and Physics*, Vol.100, No.1, (November 2006), pp. 124-127, ISSN 0254-0584.

Yoshimoto, S. Ohashi, F. & Kameyama, T. (2004), Simple Preparation of Sulfate Anion-Doped Polyaniline-Clay Nanocomposites by an Environmentally Friendly

Mechanochemical Synthesis Route, *Macromolecular Rapid Communications*, Vol.25, No.19, (October 2004), pp. 1687-1691, ISSN 1521-3927.

Zhang, L. & Dong, S. (2004), The electrocatalytic oxidation of ascorbic acid on polyaniline film synthesized in the presence of camphorsulfonic acid, Journal of Electroanalytical Chemistry, Vol.586, No.1, (July 2004), pp. 189-194, ISSN 1572-6657.

Zhang, L. (2007), The electrocatalytic oxidation of ascorbic acid on polyaniline film synthesized in the presence of β-naphthalenesulfonic acid, *Electrochimica Acta*, Vol.52, No.24, (August 2007), pp. 6969-6975, ISSN 0013-4686.

7

Electrodeposition of Polypyrrole Films: Influence of Fe_3O_4 Nanoparticles and Platinum Co-Deposition

Paula Montoya, Tiffany Marín, Jorge A. Calderón and Franklin Jaramillo
Center for Research, Innovation and Development of Materials – CIDEMAT/ University of Antioquia Colombia

1. Introduction

Inorganic particles/conductive polymers composites in bulk or films have been subject of intense study during the last decade. This type of materials offer the potential to being used in batteries, electro-chemical display devices, molecular electronics, electromagnetic shields, opto-electronic applications, microwave-absorbing materials, and even for corrosion protection (Garcia et al., 2002; McNally et al., 2005). Conducting polymers have some specific problems that make difficult its use in the above applications. Instability under oxygen and UV exposure, easily doping and over-oxidation are the most common among others. A novel strategy have been reported to improve its properties and extend the application range of these materials, this is the incorporation of inorganic particles of metallic oxides such as MnO_2, V_2O_5, TiO_2, Fe_2O_3, Fe_3O_4 and WO_3 or metallic particles of Zn, Cu, Au, Pt into the conductive polymer (Demets et al., 2000; Ferreira et al., 2001; Kawai et al., 1990; Kuwabata et al., 2000; Lenz et al., 2003; Montoya et al., 2010; Vishnuvardhan et al., 2006). For example we have recently demonstrated that the incorporation of magnetite into polypyrrole (PPy) decreases the electric resistance of the polymeric film and not only stabilize the polaronic form of the polypyrrole, but also preserve the polymer from further oxidation (Montoya et al., 2010).

Polypyrrole (PPy) exhibits interesting properties such as high conductivity, relatively good environmental stability, and wide technological applications. PPy can be obtained either by chemical and electrochemical polymerization. The electropolymerization is considered a controlled synthesis method that provides better control of thickness and morphology of films (by controlling parameters as current, voltage, and time), efficient (high material-transfer efficiency with nearly 100% material utilization and recovery), and environmentally safe (usually a water-based process). The aim of this chapter is to show in detail, two particular cases concerning the development of PPy/inorganic particles composite coatings deposited on stainless steel. First of all, a brief introduction is presented discussing the electrochemical polymerization methods. Then, as a first case, the effect of magnetite (Fe_3O_4) nanoparticles on the polymer matrix is presented. The second case is the co-deposition of platinum/PPy. Both studies show the effect of the addition of the inorganic phase on the electrical, morphological, structural and conductive properties of the polymer matrix. Finally we show the synergistic effect of PPy/Fe_3O_4/Pt composites on the final properties of the coating.

2. Electrochemical polymerization of pyrrole

Polypyrrole can be obtained by the oxidation of pyrrole monomer. Such oxidation can be accomplished by the following methods: 1) chemical polymerization in aqueous or organic media by oxidizing agents. 2) by electropolymerization on a metallic or conductive substrate by applying a potential or external current. 3) by photochemical or enzymatic catalysis polymerization. In general by chemical oxidation a final powder is obtained. Thin films are obtained by electrochemical deposition and from colloidal dispersions by enzymatic polymerization (Wallace, 2003).

Electropolymerization is a method in which a pyrrole monomer dissolved in an electrolyte solution, normally aqueous, is oxidized to form a conductive film over a work anodic electrode. Reported conductivities for electrochemically obtained PPy are between 10^{-10} – 10^{3}S/cm, due mainly to variables as deposition time, concentration of the monomer, substrate and deposition method. Figure 1 shows a simplified mechanism of oxidation of PPy, where n can be 3-4 and m is related with the chain length and determining the molecular weight. A^- is a counter ion from the electrolyte to balance the charge over the polymer.

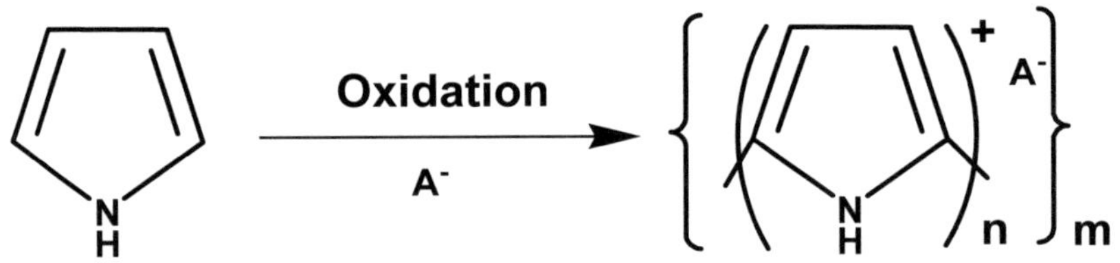

Fig. 1. Simplified mechanism of the polymerization of polypyrrole.

2.1 Steady state methods

These methods consist in applying a constant potential or current to an electrode giving place to a constant response upon time in current or potential, respectively. Between two, potentiostatic method consist in fixing the work electrode potential to give as a result the curves illustrated in Figure 2. The galvanostatic method is the inverse to that mentioned above. Steady state methods allow following electrode processes and changes in a system. This can be achieved by recording an electrical parameter upon time whereas the other one is being fixed, they can be: 1) Potentiostatic, in which a potential pulse is applied to the electrode and the current is registered as a function of time. During the experiment, once the double layer is charged, the potential of the electrode remains at a constant value (E_a) over the open circuit voltage (E_{ocp}) and the current decreases as the concentration of the electrodic specie decreases in the solution and is deposited in the working electrode. The described events are depicted in Figure 2. 2) Galvanostatic, when a constant current pulse is applied to the work electrode, the potential is shifted from the equilibrium and changes are registered against time.

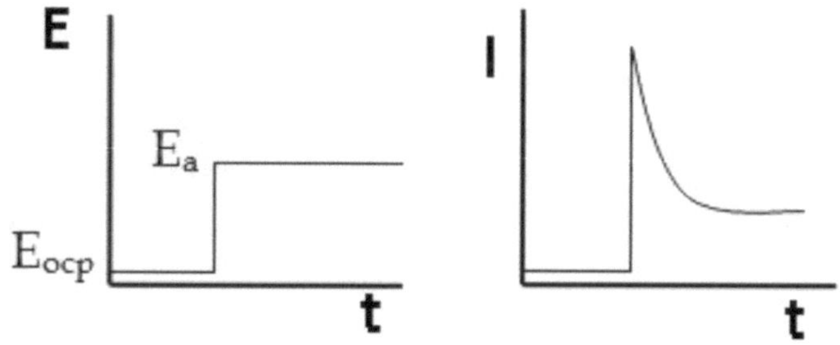

Fig. 2. Potentiostatic polarization and the current response of a typical deposition process of PPy.

2.2 Unsteady state methods

Unsteady state methods allow following electrode processes and changes in a system that do not occur instantaneously. Between non-stationary methods available for studying electrode processes, potential sweep methods are probably the most widely used. They consist in the application of a continuously time-varying potential to the working electrode. The observed current is therefore different from that in the steady state. Resulting this in the occurrence of oxidation or reduction reactions of electroactive species in solution (faradaic reactions), possibly adsorption of species according to the potential, and a capacitive current due to double layer charging. There are two forms of sweep voltammetry techniques named, linear sweep voltammetry (LSV) and cyclic voltammetry (CV). In linear sweep voltammetry the potential is scanned only in one direction, stopping at a chosen value, E_{fin} for example at $t = t_1$ in Figure 3.

The scan direction can be positive or negative and, in principle, the sweep rate can have any value. In cyclic voltammetry, the potential scan is done in two directions, on reaching $t = t_1$ the sweep direction is inverted as shown in Figure 3 and swept until E_{min}, then inverted at t_2 and swept to E_{max}. The faradaic current, I_f, due to the electrode reaction, is registered simultaneously to the applied potential where electrode reactions occur, giving place to a cyclic voltammogram like that shown in Figure 4.

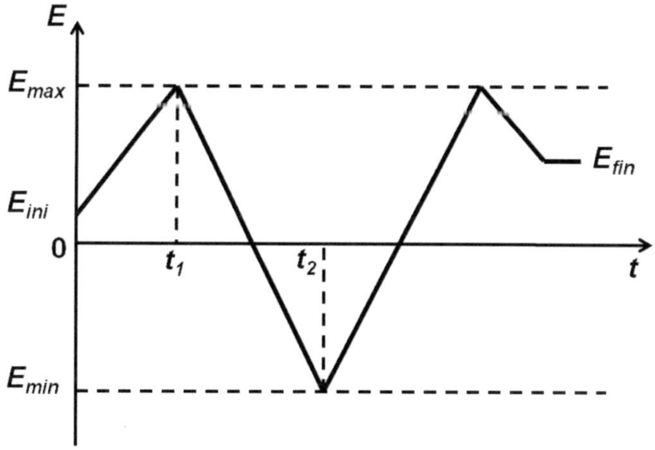

Fig. 3. Variation of applied potential with time in cyclic voltammetry, showing the initial potential, E_{ini}, the final potential, E_{fin}, maximum, E_{max}, and minimum, E_{min}, potentials.

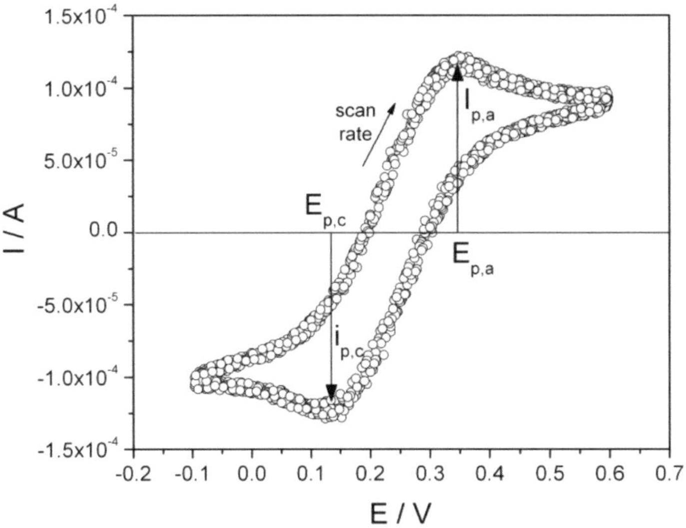

Fig. 4. Typical Cyclic voltammogram for a reversible system.

3. Polypyrrole/Fe_3O_4 composite films

3.1 General issues Polypyrrole/Fe_3O_4 films

Several studies have attempted to the development of PPy/Fe_3O_4 composite films; however there is not still a general consensus about the effect of magnetite on the final composite as well as the type of interaction between the PPy and Fe_3O_4 nanoparticles. Some researchers have found that the presence of Fe_3O_4 in the PPy causes an increase in the conductivity of the composite as compared with pure PPy (Chen et al., 2004; Chen et al., 2003). However, there are others authors claiming the opposite (Deng et al., 2003a; Pailleret et al., 2007). The controversy may be due to many factors such as oxide particle size, dopant agent, electrolyte, method of synthesis, electrodeposition rate, etc.; which influence the properties of the composite leading to changes in the electrical properties when the amount of magnetite incorporated into the polymer matrix is increased. Composites of organic conducting polymer and Fe_3O_4 particles have the advantage of having both good electrical and magnetic properties. In addition of being promising materials as protective coatings, they are attractive to be used in new batteries, fuel and solar cells, capacitors or magnetic materials.

3.2 Electrochemical polymerization of the films

The production of adherent coatings based on conductive polymers chemically synthesized is hampered by its low solubility in common solvents. Polymers obtained by this method cannot be processed by spin or dip coating. This disadvantage can be overcome by electrochemical polymerization, which can simultaneously allow to form and to deposit polymer coatings on the substrate from a monomer–electrolyte solution. Unfortunately, as shown by several previous studies, the attempts to electrochemically polymerize pyrrole on reactive metals such as iron or other oxidizable metals in aqueous medium, present some difficulties. In this case is necessary to find the electrochemical conditions that lead to a partial passivation of the metal and decrease its dissolution rate without avoiding the electropolymerization of the monomer. In acidic medium, no noble metals are preferentially dissolved due to the polymerization potential of pyrrole, being this higher than the oxidation potential of the metal. These metals can be covered with passive films that prevent their dissolution in acidic medium. However, these passive films are either soluble in the polymerization medium or poorly wetted by the electrolyte-monomer solution and thereby preventing electropolymerization of the monomer. Therefore a coating process that can force simultaneous passivation and polymerization is desirable. Some researchers have developed a one-step in situ passivation and coating of steel for polypyrrole with good adhesion properties (Iroh and Su, 2000, 2002; Ocon et al., 2005; Wencheng and Iroh, 1998).

In the same way, PPy/iron oxides composites can be prepared by electrochemical oxidation of the monomer in the presence of dopant anion and the iron oxide particles in suspension. Some works have reported composite films obtained by galvanostatic method on several substrates as: iron, stainless steel, carbon steel, platinum, etc. (Garcia et al., 2002; Montoya et al., 2010; Wencheng and Iroh, 1998). Instead, other electrochemical techniques as constant or variable electric potential have been imposed in order to synthesize the composite coatings. Independently of the electrochemical technique used to film formation the mechanism of electrosynthesis is the same and it involves different stages, as showed in Figure 5: 1. Monomer oxidation. 2. Radical-radical coupling. 3. Deprotonation/Re-aromatization and 4. Propagation or subsequent oxidation. The polymerization is believed to proceed via a radical-radical mechanism (Andrieux et al., 1991), wherein the natural repulsion of the radicals is supposed to be refused by the solvent, the counterion, and even the monomer. Chain growth then continues until the charge on the chain is such that a counterion is able to be incorporated. The backbone has a delocalized π-system and the polymer film incorporate dopant anions, stabilizing the charge on the backbone of the polymer (Ocon et al., 2005). Eventually, as the polymer chain exceeds a critical length, the solubility limit is exceeded, and the polymer can be deposited on the electrode surface. However, in the early stages of polymerization the electrode substrate plays a critical role once the reaction is initiated. For example, when the PPy is electropolymerized on carbon steel by galvanostatic method, using oxalic acid as the electrolyte, three stages can be distinguished in the process of electropolymerization, as can be seen in Figure 6. In the first stage the dissolution of steel is observed, followed by its passivation by the formation of iron oxalate film and iron oxides, and finally the deposition of polypyrrole. Passivation time can be decreased by increasing the applied current, the increased pH and the concentration of pyrrole (Su and Iroh, 1997). Earliest results show that the passive layer has a well-defined crystalline structure, insoluble in water enough to encourage further electropolymerization. It is appropriate to emphasize the formation of iron/oxalic acid system, principally in terms of its complexing and salt-forming reactions (Giacomelli et al., 2004). It is worth noting that the process of electropolymerization of pyrrole on stainless steel is similar to carbon steel if the passive layer of stainless steel is removed prior to the electrosynthesis, otherwise, the time required to start the electropolymerization of the monomer is almost negligible, see Figure 7. During the electrodeposition of PPy and PPy/Fe_3O_4 coatings on stainless steel is possible to see some changes in the registered potentiometric curves. The Figure 7 shows the curves E vs. t for the electropolymerization of pure PPy and PPy/Fe_3O_4 composite films with different concentrations of magnetite present in the electrolyte of the synthesis. It can be observed that the increasing of magnetite content in the polymer decrease the electropolymerization potential. This could indicate that the presence of magnetite reduces the required energy for polymer film formation and prevents the overoxidation of the film during the polymerization process (Montoya et al., 2009).

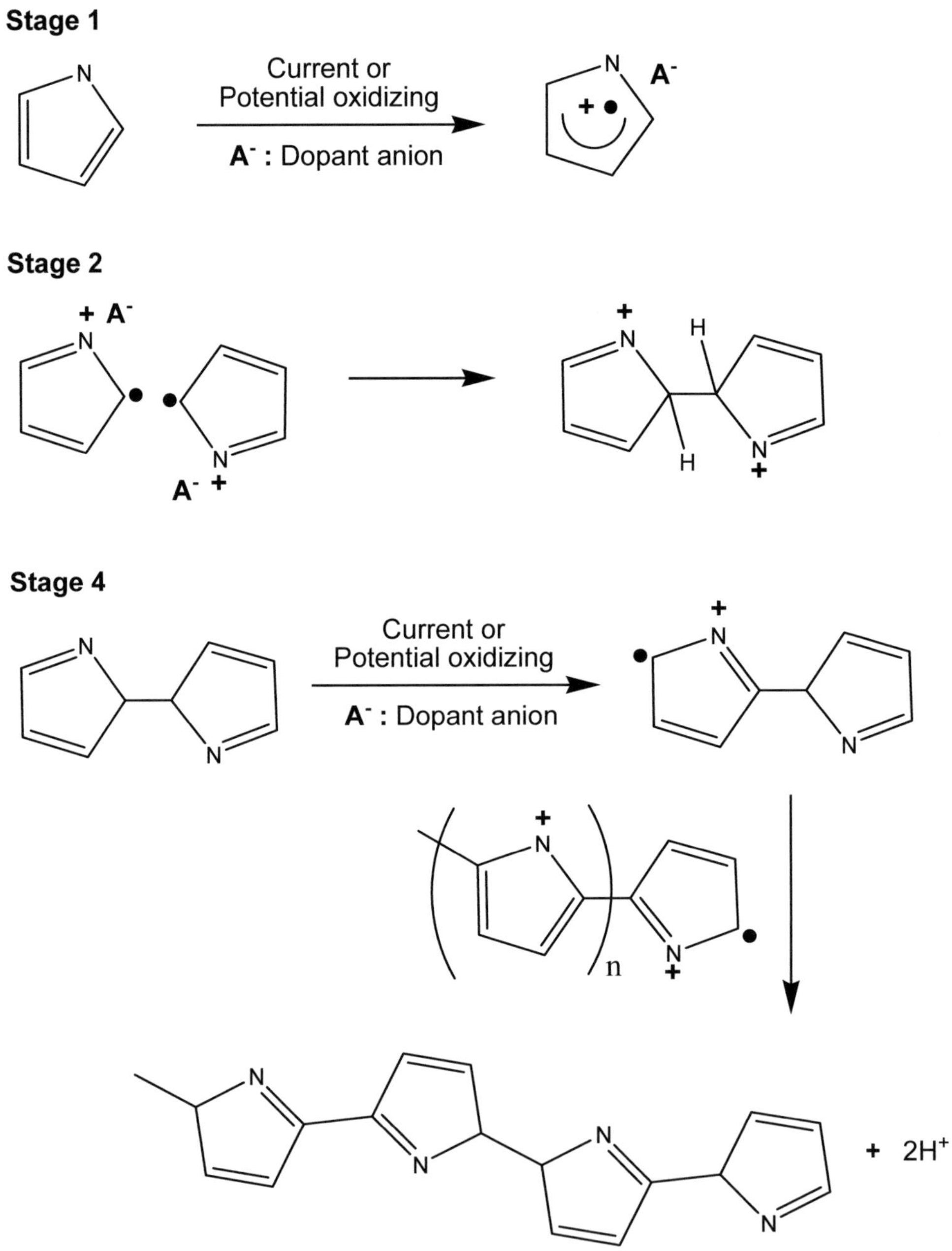

Fig. 5. Mechanism of formation of polypyrrole by electropolymerization

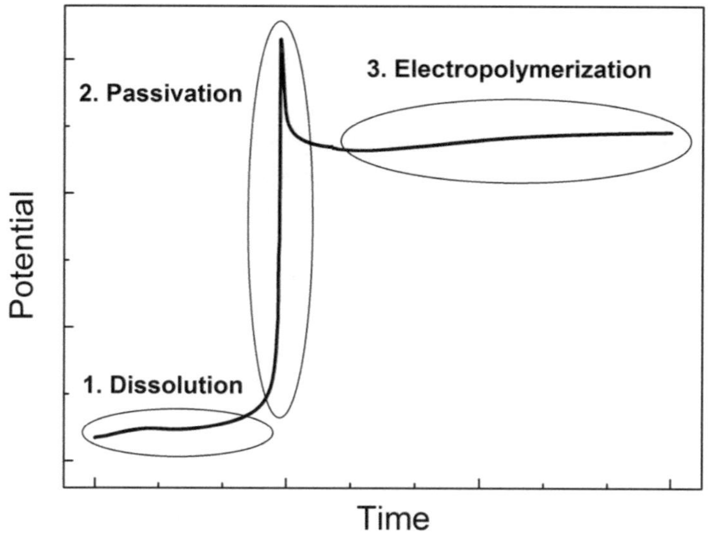

Fig. 6. Potentiometric curve during the formation of Polypyrrole/Fe_3O_4 coatings on carbon steel in oxalic acid by galvanostatic method.

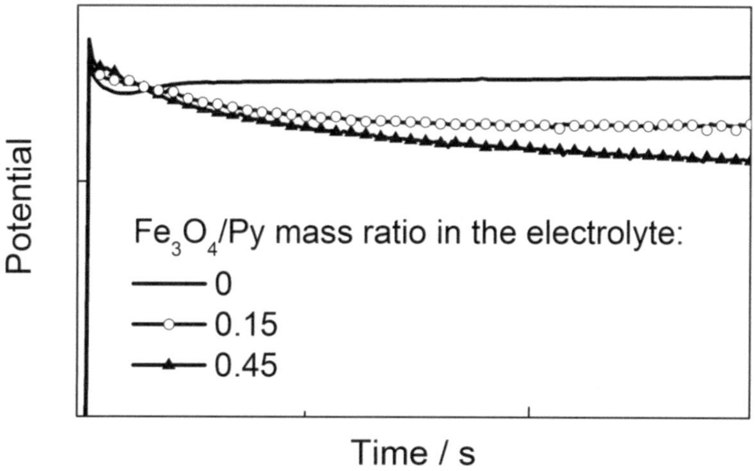

Fig. 7. Potentiometric curve during the galvanostatic electrodeposition of PPy/Fe_3O_4 coatings on stainless steel with several Fe_3O_4/Py mass ratios in the electrolyte of synthesis.

3.3 Physical chemical interaction and stabilization of the polymer

Several studies have shown the interactions between polypyrrole and Fe_3O_4 in polypyrrole-Fe_3O_4 composite films. A recent study examined the effect of the magnetite particles on the structure of the polymer matrix. Films of polypyrrole were synthesized on stainless steel in the presence of magnetite particles (Montoya et al., 2010). The effect of the magnetite particles on the structure of the polymer matrix was determined using Raman spectroscopy and Scanning Electron Microscopy (SEM). Additionally, the changes in the electrical resistance of the films were evaluated over time by electrochemical impedance spectroscopy in solid state. These results showed that the interaction between the incorporated magnetite and the polymeric matrix, leads to morphological and electronic changes in the composite film respect to pure PPy. It is possible to see in the micrograph of the Figure 8 that the polymer is preferentially deposited as globular chains along the lines of the polished surface. As the content of Fe_3O_4 increases in the polymer matrix, it is possible to observe a change in the coating morphology going from agglomerated globular structures composed of average size about 100nm (Figure 8a) to agglomerates with a star-like conformation (Figure 8b). These results clearly indicate the effect of Fe_3O_4 particles in the morphology of the polymer.

Figure 9 shows the Raman spectra obtained for pure PPy film and Fe_3O_4/Py composite film with a mass ratio in the electrolytic solution of 0.75. The spectra present the characteristic bands already attributed to PPy vibrational modes located at 1592, 1382, 1320, 1252, 1082, 1047, 982 and 931cm^{-1} (Santos et al., 2007). By comparing with reported data by in-situ Raman studies (Furukawa et al., 1988; Santos et al., 2007), it is clear that the polymer is in the oxidized state where both aromatic and quinoid structures coexist indicating a pseudo-equilibrium between polaronic and bipolaronic structures at the same oxidation level as already pointed out by Santos (Santos et al., 2007) and Furukawa (Furukawa et al., 1988). Figure 9 shows that the intensity of the band at 931cm^{-1} decreases when the Fe_3O_4 is in the film. This band is characteristic of the bipolaronic structure (quinoid form) in the oxidized state and is attributed to out-of-plane C-H deformation. Additionally, two bands characteristic of the polaronic structure (aromatic form), at 982 and 1047cm^{-1}, assigned to ring deformation mode and CH in-plane deformation mode respectively, increased when the Fe_3O_4 is present in the composite (Furukawa et al., 1988). Both the bands at 1500 and 982cm^{-1} are assignable to aromatic segments but these are also present at the most oxidized state. The band at 1082cm^{-1}, presents in both spectra is assigned to the most oxidized state, correlated to the bipolaron or dication state. The intensity of this band remains relatively constant for all the obtained composites. According to Yakushi et al. (Yakushi et al., 1983), the evidence of aromatic segments bands when the polymer is oxidized is due to the presence of different conjugation lengths in the PPy films, which are also observed in the composite PPy–Fe_3O_4.

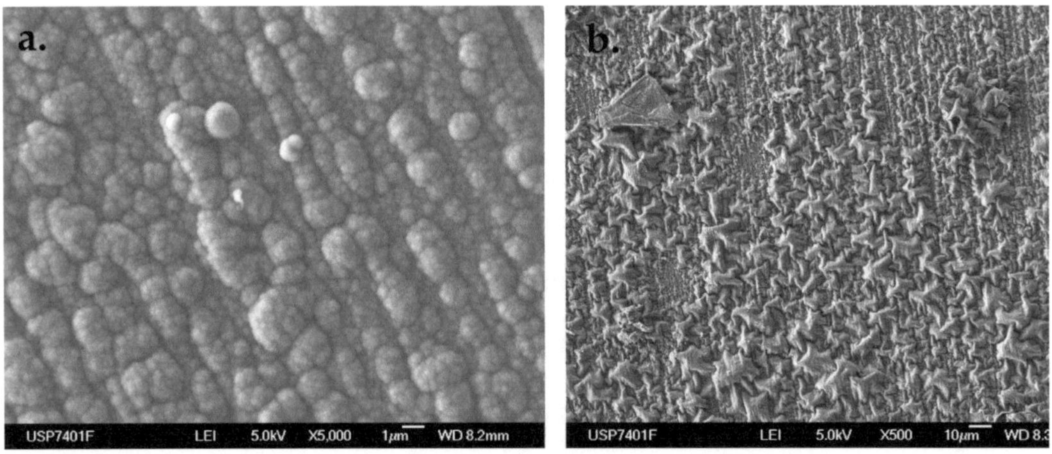

Fig. 8. SEM images of a. pure PPy film and b. Fe_3O_4/Py composite film with a mass ratio in the electrolytic solution of 0.75.

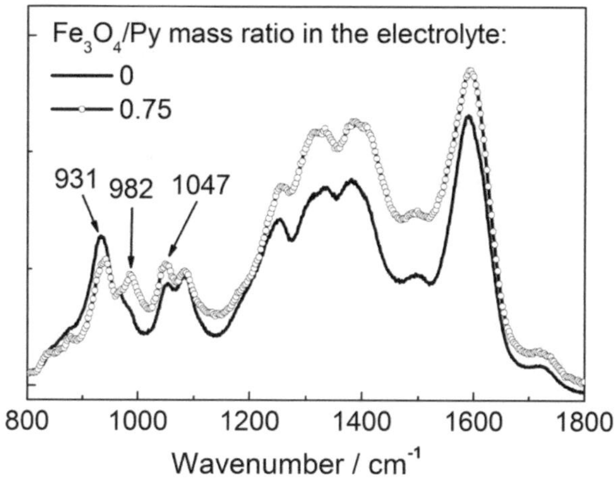

Fig. 9. Raman spectra of PPy film and composite film with a Fe_3O_4/Py mass ratio in the electrolytic solution of 0.75.

The most important evidences in this work are as follows. The incorporation of magnetite into polymeric matrix decreases the electrical resistance of the polymeric films. Additionally, the presence of magnetite into the film not only stabilizes the polaronic form of the polypyrrole, but also preserves the polymer from further oxidation. These results were consistent with previous results showed by Elliot et al.(Elliott et al., 1991); the authors combined resistance and electron spin resonance (ESR) measurements of PPy films and they observed that electric resistance is minimum when the ESR signal is maximum, indicating that polaronic species are the main responsible by charge transportation (Elliott et al., 1991). In the same direction, investigations performed by Deng et al. suggested that the interaction mechanism between the PPy and the Fe_3O_4 may involve the interaction between lone pair electrons of the nitrogen atom in PPy chain with the 3d orbital of iron atom to form a coordinate bond, reducing the energy level interval of the pyrrole ring (Deng et al., 2003). Additionally, the work of Tzong-Ming et al. also shows evidences of the interaction between PPy and Fe_3O_4 nanoparticles (Wu et al., 2007). They observed by UV spectroscopy, a transition from the valence band to the antibonding polaron state in PPy, indicating that the produced PPy was in the doped state. When the spherical structure of PPy-Fe_3O_4 nanocomposites was formed, the characteristic peak assigned to the polaron-p transition was slightly shifted to a smaller wavelength with increasing Fe_3O_4, suggesting the interaction between the quinoid rings of PPy and Fe_3O_4. This is also consistent with the results observed for us by Raman spectroscopy in a recent study (Montoya et al., 2010) and also with the fact that magnetite preserves the PPy film to further oxidation and the PPy-Fe_3O_4 could preserve the conductive state for longer time due to the stabilization of the polaronic form of the oxidized state.

4. Polypyrrole/Pt composite films

4.1 Polypyrrole/Pt films general description

Composite materials based on conducting polymers are alternatives with equal or better efficiency that metals typically used in fuel cell electrodes. These types of materials are considered one the most promising group of polymers to manufacture these cells because their easy application and operation, being PPy one of them. In recent years it has been carried out the electrodeposition of PPy with the incorporation of platinum, silver or gold on substrates such as glassy carbon, gold and platinum. The electrochemical synthesis of PPy/Pt films and their morphological and structural properties are presented here (Marín et al., 2010; Marín et al., 2009). The results in this section show the formation of sandwich-like structures and the important stabilization effect that platinum provided to the polymeric film in terms of electroactive activity. The Polypyrrole/Pt films are a proposed system based on two configurations: two-layer and multiple-layer. The films were obtained by electropolymerization of pyrrole and cathodic deposition of platinum from ammonium hexachloroplatinate salt on stainless steel 304. Moreover, to improve PPy/Pt films, a multiple-layer configuration was obtained by using an electrolytic solution containing pyrrole and platinum salt. Finally the electrocatalytic activity of two-layer and multiple-layer PPy/Pt films was evaluated for methanol oxidation.

4.2 Electrochemical polymerization

The experiments are classified into two groups, two-layer and multiple-layer, as shown in Table 1 and Figure 10. Ammonium Hexachloroplatinate was obtained by hydrometallurgical processes using alluvial platinum (Benner, 1991; Georgieva and Andonovski, 2003) and used as Pt source for the electrochemical synthesis. All potentials reported in this chapter are referred to calomel electrode.

Two-layer PPy/Pt	1. Electrochemical synthesis of PPy films 2. Reduction of Pt (IV) by a potentiostatic method.
Multiple-layer PPy/Pt	Monomer oxidation and platinum reduction alternating.

Table 1. Description of the experiments.

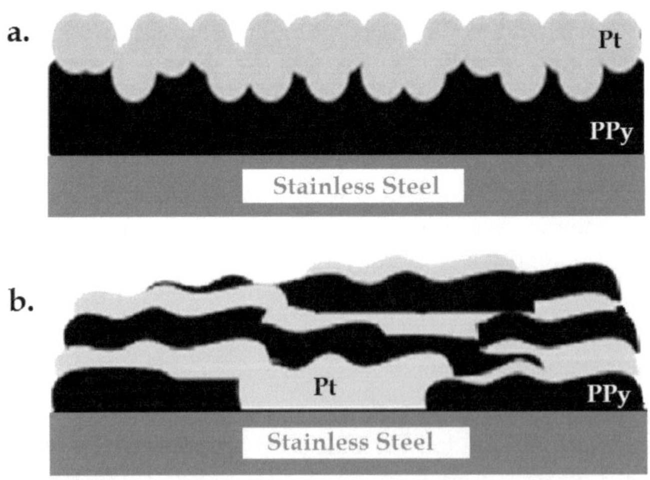

Fig. 10. Schematic representation of PPy composite films a. two-layer, b. multiple-layer.

In the first film configuration the PPy films were obtained by a galvanostatic method, applying 0.001A for 2400s using a 0.1M pyrrole + 0.1M H_2SO_4 electrolyte solution. Once obtained the PPy film, a layer of platinum was cathodically electrodeposited on PPy film in order to obtain the two-layer PPy/Pt (1). The Pt layer was obtained by potentiostatic method from solution containing 0.001M of $[(NH_4)_2PtCl_6]$ and 0.5M of H_2SO_4 and applying -0.2V during 1200s. The multiple-layer PPy/Pt (2) composite films were obtained by potentiostatic method applying 0.5V to get PPy films, and then applying -0.2V to get the platinum deposition using a solution containing 0.1M of Pyrrole, 0.001M of $[(NH_4)_2PtCl_6]$ and 0.5M of H_2SO_4. Each potential was applied during 1200s.

4.3 Characterization

4.3.1 Scanning electron microscopy (SEM)

Figure 11a illustrates how the polymer was formed along the lines of the substrate, showing a globular morphology of sizes ranging from 4 to 5μm with a few smaller outer cores on the surface, which are highlighted on the entire matrix of the polymer. This globular morphology, as previous reported, is typical of polypyrrole when formed by electropolymerization either by potentiostatic or galvanostatic methods (Lehr and Saidman, 2007; Lu et al., 2006; Martins et al., 2008; Nie et al., 2008). In Figures 11b and 11c the same globular morphology is observed. However, in both films there are not small outer cores as in pristine PPy. It is then evident that platinum homogeneously covered the PPy film and copied its morphology. The obtained films were evaluated with EDS analysis to confirm the

Fig. 11. SEM micrographs of films: a. pure PPy, b. two-layer PPy/Pt, c. multi-layer PPy/Pt.

presence of platinum. Figure 12 shows the results of this analysis. The atomic percentage of Pt is higher than that found for other elements. Fe, Ni and Cr are elements associated with the stainless steel substrate. It is observed that for multiple-layer films, elements of the substrate are in less percentage, given the increase in thickness. In both films it is evident the presence of platinum on the surface of the polymer matrix.

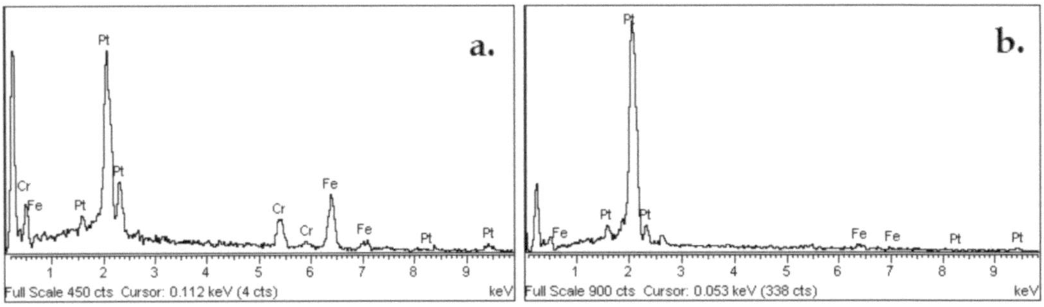

Fig. 12. EDS spectra: a. two-layer PPy/Pt , b. multi-layer PPy/Pt.

Figure 13a shows a SEM micrograph taken at a cross section of the two-layer PPy/Pt film. In this micrograph two layers were clearly observed. The first layer is the PPy and second layer correspond to Pt. The Figure 13b represents a top view of the multi-layer PPy/Pt film. The micrograph shows the effect of electrical potential alternation during the films deposition in short times, giving as a result the formation of discontinuous regions covered by the polymer (black regions) and platinum (bright regions), without being able to cover the entire surface of the substrate. Finally it is expected that both films presented electrocatalytic activity (Bouzek et al., 2001, b; Holzhauser et al., 2005). However, the two-layer films can additionally act as a protective coating for stainless steel substrate, because complete coverage of the metal substrate by PPy.

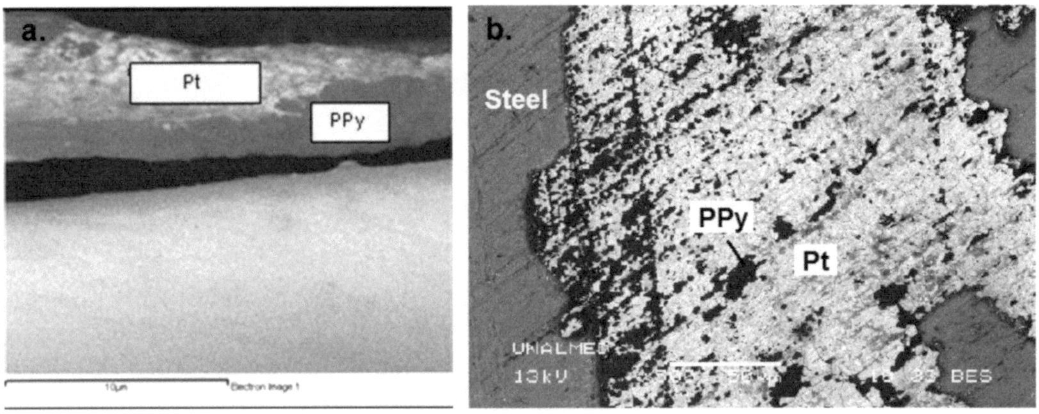

Fig. 13. SEM micrographs of: a. cross-sectional view of two-layer PPy/Pt films, b. top view of multiple-layer PPy/Pt films.

4.3.2 Atomic force microscopy (AFM)

All the obtained films were analyzed by AFM to study their homogeneity, the grain size and surface roughness. The Figure 14a shows a PPy image with granular morphology heterogeneously distributed all over the surface. Some authors agree that this is the morphology adopted by PPy formed on a metal substrate (Bravo-Grimaldo et al., 2007; Okner et al., 2007). An average grain size is 0.125µm and a surface roughness of 33.6nm could be calculated, similar to that observed by J. Tamm (Tamm et al., 2004). In the Figure 14b the morphology of a typical two-layer PPy/Pt film is observed. That film showed similar morphology to that for PPy. This is an indication that platinum covers much of the polymer matrix and takes its morphology. Pt reduced the grain size up to 0.083µm. The surface roughness was less modified and a value of 33.3nm was calculated for this film. Figure 14c shows the morphology for multiple-layer PPy/Pt films and as the two-layer films, they have a granular morphology and platinum copies the polymer. Similar grain size values compared to two-layer film were obtained, and an increase of surface roughness of 10nm was observed.

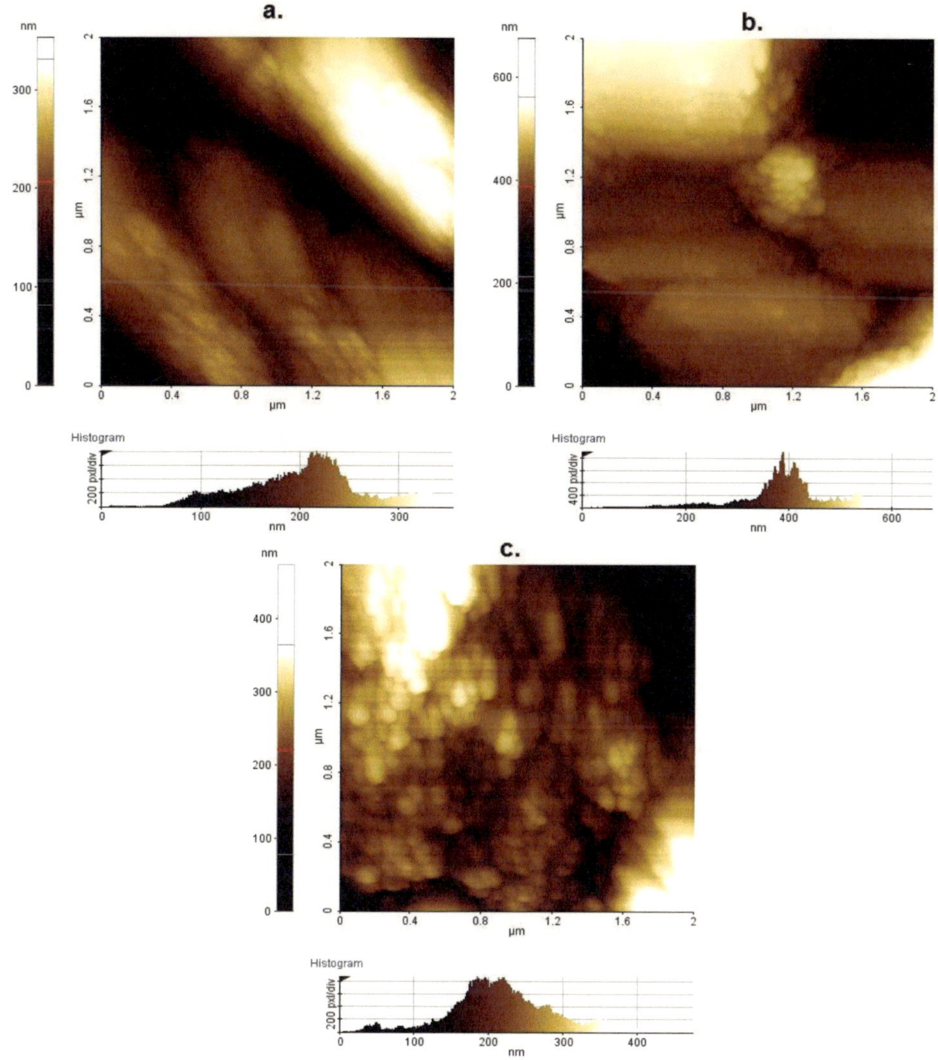

Fig. 14. AFM Image: a. PPy, b. two-layer PPy/Pt, c. multiple-layer PPy/Pt films.

4.3.3 Surface profilometry

This technique was used to obtain accurate results of film thickness. Table 2 reports the thickness data for each system. The obtained results are initially consistent with the patterns shown by Figure 10. The platinum coating deposited on the PPy film had a thickness of about 0.14 μm.

Film	Thickness (μm)
PPy	1.28
Two-layer PPy/Pt	1.42
Multiple-layer PPy/Pt	1.98

Table 2. Thickness of the films.

4.4 Fuel cells applications

Polypyrrole/Pt composite films were evaluated by cyclic voltammetry as anodes for fuel cells applications. The ability of the films to oxidize methanol was measured in 0.5M of H_2SO_4 and 1M of MeOH (Bouzek et al., 2001a; Holzhauser et al., 2005). The cyclic voltammetry curves were run without going through the over oxidation potential of the polymer, so these were taken from the open circuit potential to 0.8V in the anodic direction and to -0.5V in the cathodic direction.

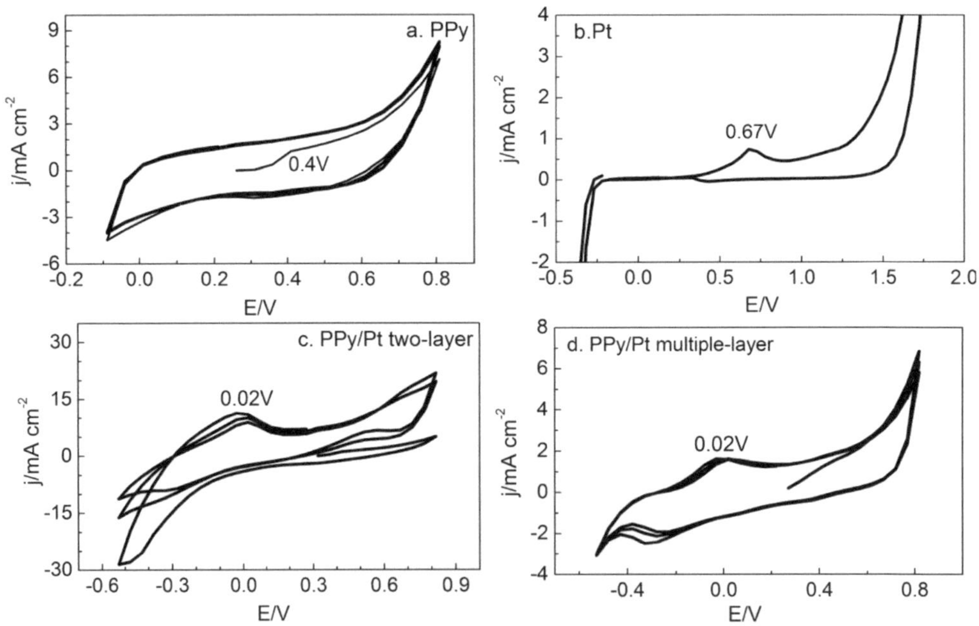

Fig. 15. Cyclic voltammograms of: a. PPy, b. Pt, c. two-layer PPy/Pt and d. multiple-layer PPy/Pt in 0.5 mol.L^{-1} H_2SO_4 and 1 mol.L^{-1} methanol at 100 mV/s and atmosphere of N_2.

Although in both, two-layer and multiple-layer PPy/Pt films, the methanol oxidation occurs at 0.02V, the kinetics of methanol oxidation is different. This is mainly due to the fact that two-layer films presented higher oxidation current densities (11mA.cm^{-2}) than those observed for multiple-layer films ($2mA.cm^{-2}$). The results show that two-layer films are more efficient to oxidize methanol because the more active layer of platinum is deposited on all the surface of the PPy film. While in the multiple-layer films the platinum is deposited heterogeneously all through the film.

5. PPy/Fe_3O_4/Pt composite films

In the present section the effect of the concentration of H_2SO_4 as the electrolyte used in the synthesis of PPy, PPy/Pt, PPy/Fe_3O_4, and PPy/Fe_3O_4/Pt films deposited on 304 stainless steel is evaluated. The oxidation potentials of all films were determined by cyclic voltammetry technique in order to assess the effect of the incorporation of Pt and/or Fe_3O_4 into the conductive polymer matrix. The inorganic particles significantly influence both the oxidation stability and morphology of the PPy matrix. The results showed that the incorporation of platinum and magnetite nanoparticles into the polypyrrole film increase the current density of the electrochemical response and shift the oxidation potential towards more anodic values. Moreover, the deposition of platinum and magnetite allows larger over-potential window to promote redox reactions without compromising the stability to oxidation of the polymer.

5.1 Electrochemical polymerization

Ammonium Hexachloroplatinate was obtained by hydrometallurgical processes of from alluvial platinum (Benner et al., 1991; Georgieva and Andonovski, 2003). This salt was used as Pt source for the electrochemical synthesis. The magnetite particles were obtained by hydrothermal method reported in the literature with some modifications (Deng et al., 2003b; Huang et al., 2005).

The polypyrrole films were obtained by galvanostatic method applying 1mA during 2000s. 0.1M of pyrrole and 0.1M of H_2SO_4 or 0.25M of H_2SO_4 were used as electrolyte. The PPy/Fe_3O_4 films were obtained by galvanostatic method with the same conditions as for PPy films. Before the electropolymerization, Fe_3O_4 particles were dispersed into the electrolyte solution, using an ultrasonic probe during 30 minutes. The PPy/Pt and PPy/Fe_3O_4/Pt composite films were made in two stages, each one in a separate experiment. In the first stage a PPy or PPy/Fe_3O_4 film was obtained by galvanostatic method. The platinum was cathodically electrodeposited on PPy or PPy/Fe_3O_4, from 0.001M of $[(NH_4)_2PtCl_6]$ + 0.5M of H_2SO_4 solution by potentiostatic method applying -0.2V during 1200s. The experiments were classified as shown in Table 3.

Film	[H_2SO_4]	Test
PPy **PPy/Pt** **PPy/Fe_3O_4** **PPy/Fe_3O_4/Pt**	**0.1M**	**(1)**
PPy **PPy/Pt** **PPy/Fe_3O_4** **PPy/Fe_3O_4/Pt**	**0.25M**	**(2)**

Table 3. Classification of experiments according to the concentration of the acid used as electrolyte in the polymerization.

5.2 Characterization

5.2.1 Scanning electron microscopy (SEM)

The effect of the electrolyte concentration in the morphology of PPy films was evaluated. In Figure 16 the resulting morphologies of PPy in the experiments (1) and (2) are presented.

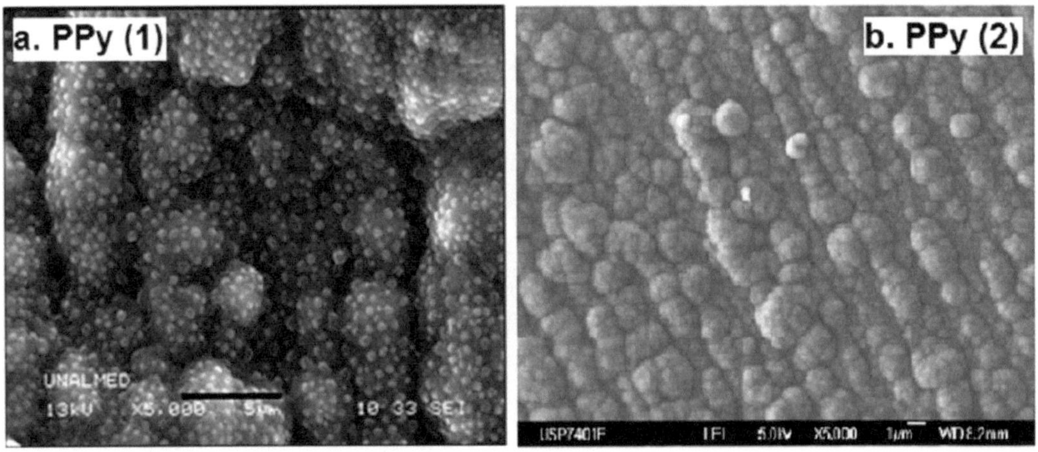

Fig. 16. SEM micrographs of the films: a. PPy (1), b. PPy (2).

Figure 16a shows polypyrrole obtained in 0.1M of H_2SO_4 named as PPy (1). The typical globular morphology is observed (Lehr and Saidman, 2007; Martins et al., 2008). The Figure 16b presents a polypyrrole obtained in 0.25M of H_2SO_4, PPy (2). The influence of increasing the electrolyte concentration can be clearly observed. A smaller size of about 100 nm was found for the smallest domain. Moreover these films did not show the formation of outer core on the surface.

5.2.2 Cyclic voltammetry

The effect of Fe_3O_4 and Pt particles incorporated into the polypyrrole matrix were evaluated by cyclic voltammetry technique. The oxidation potential of PPy film was analyzed in order to assess the effect of inorganic particles into the film. Figures 17, 18, 19, and 20 shows the PPy, PPy/Pt, PPy/Fe_3O_4 and PPy/Fe_3O_4/Pt cyclic voltammograms evaluated in 0.5M of H_2SO_4 as electrolyte solution for each of the concentrations in which the polymer was obtained, experiments (1) and (2). Figure 17 is composed by three curves. The first one shows the response from stainless steel giving two oxidation current peaks at -0.3V and -1.4 V being the last one the most intense. The second current peak also occurs during the sweep return, resulting in an anodic reactivation process of the substrate. The second and third voltammetry curves show PPy films obtained on stainless steel in solution of 0.1M and 0.25M of H_2SO_4, respectively. The over oxidation potentials of polypyrrole appeared at 0.9V and 1.1V. These oxidation current peaks are responsible of the enhance presence of bipolarons rather than polarons in the polymer chain (Lamprakopoulos et al., 2004; Radhakrishnan and Adhikari, 2006)

The over oxidation potential of polypyrrole is not influenced by the electrolyte concentration, oxidation current peaks are observed at the same potential in PPy films obtained with both electrolyte concentrations. However the current densities in these potentials are greater for PPy films obtained at 0.25M of H_2SO_4. This means that the chemical doping is higher for the PPy (2) at higher electrolyte concentration than PPy (1) obtained at a lower concentration. The sweep return in cyclic voltammetry curves in the PPy films presented a reactivation peak at 1.4V due to the substrate, showing that after the over oxidation potential of the film the steel is exposed to an aggressive media.

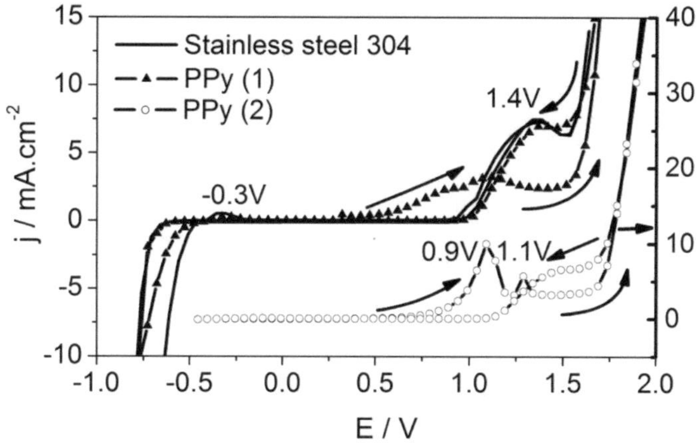

Fig. 17. Cyclic voltammograms of PPy films in the experiments (1) - left axis and (2) – right axis in 0.5 mol.L^{-1} H_2SO_4 at 5 mV/s.

The Figure 18 shows cyclic voltammograms of PPy/Pt (1) and PPy/Pt (2) composite films in 0.5M of H_2SO_4 electrolyte. Both films presented two oxidation peaks. The first peak in the anodic direction at 0.9V is an over oxidation of the polymer matrix. The second peak in the cathodic direction at 1.4V corresponds to the substrate oxidation process. When comparing the voltammograms in Figures 17 and 18 it can be concluded that although platinum is able to cover all the electrodeposited PPy film, this does not inhibit the oxidation of the polymer. It was observed that both samples of PPy/Pt (1) and (2) presented an over oxidation potential at 0.9V and current density decreases in this potential for PPy (2) film.

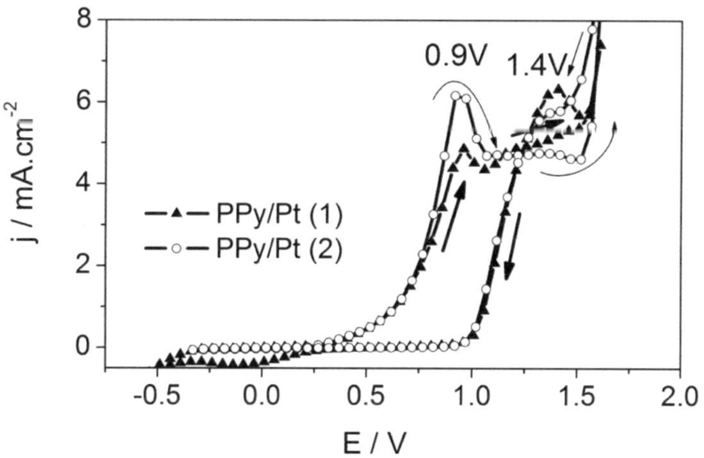

Fig. 18. Cyclic voltammograms of PPy/Pt films in the experiments (1) and (2) in 0.5 mol L^{-1} H_2SO_4 at 5 mV/s.

Figure 19 shows the voltammograms of PPy/Fe_3O_4 composite films. There are some differences compared to voltammetry response of PPy/Pt films. The first difference is that the oxidation potential of PPy film is shifted to more anodic potentials going from 0.9V to 1.25V for PPy/Fe_3O_4 composite film obtained in 0.1M of H_2SO_4 and from 0.9V to 1.36V for PPy/Fe_3O_4 composite film obtained in 0.25M of H_2SO_4. The observed shift could be associated with a possible redox effect that magnetite provides to overoxidation of the polymer as demonstrated previously (Garcia et al., 2002; Montoya et al., 2009). The second event is an increase of the current density related to polymer oxidation, being considerably higher when the composite film is obtained in an electrolyte solution with high concentration of counter-ions as well as was observed for the PPy and PPy/Pt films. As expected, the PPy/Fe_3O_4 composite film obtained in 0.25M of H_2SO_4 shows greater doping degree of counter-ions than the PPy/Fe_3O_4 composite film obtained in 0.1M of H_2SO_4. From voltammograms of Figure 17 one current peak at 0.4V can be possibly associated to magnetite oxidation. The peak observed for the cathodic sweep at 1.4V was associated with substrate oxidation due to debonding of the film.

Figure 20 shows the voltammograms of PPy/Fe_3O_4/Pt composites films for experiments (1) and (2). In these curves is clearly observed the new positive effect of magnetite due to a shift of the over oxidation potential of PPy to a more cathodic potential at 1.28V. On the other hand, the curves here showed oxidation current densities of the polymer higher than that observed for PPy films. The composite system showed a slightly difference when the electrolyte concentration increased, as observed in the Figure 20. Hence, is likely that the platinum completely covers the surface of the polymer and decreases the counter-ions diffusion through the film when subjected to oxidation-reduction cycles.

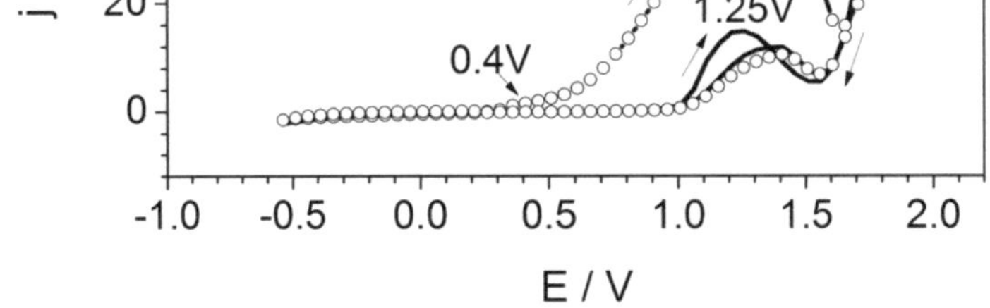

Fig. 19. Cyclic voltammograms of PPy/Fe_3O_4 films in the experiments (1) and (2) in 0.5 mol L^{-1} H_2SO_4 at 5 mV/s.

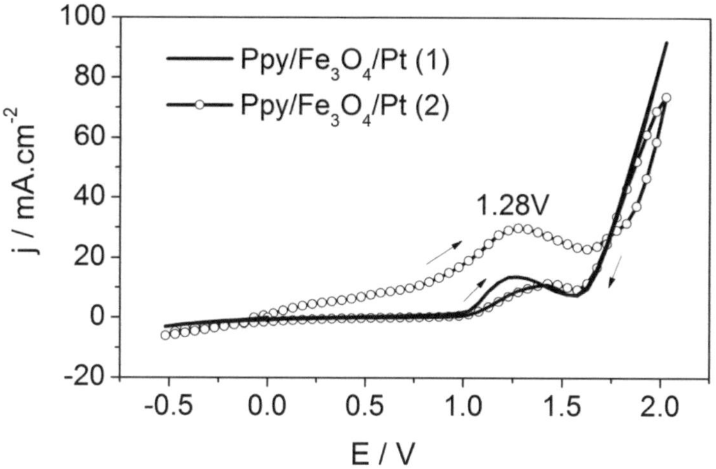

Fig. 20. Cyclic voltammograms of PPy/Fe_3O_4/Pt films in the experiments (1) and (2) in 0.5 mol L^{-1} H_2SO_4 at 5 mV/s.

6. Conclusions

The addition of Fe_3O_4 to polypyrrole matrix modifies the final structure of the polymer films, as evidenced by SEM and Raman spectroscopy. The observed changes in morphology and structural vibrations of the polymeric matrix after particle incorporation indicate that magnetite interacts with polypyrrole and modifies its properties. The charge transfer process between PPy and Fe_3O_4 causes the stabilization of polaronic segments in decreases the bipolaronic states. This fact occurs in a diminution of the electric resistance of the composite upon Fe_3O_4 load and the preservation of the conductivity when the film is exposed to moisture. The presence of magnetite reduces the required energy for the polymerization and prevents over-oxidation of the polymer during the film formation.

Incorporation of platinum into PPy films does not modify the chemical structure of polymer, however increases the electric conductivity of the composite film. When Pt is incorporated into PPy film the oxidation potential of the polymer is shifted to more positive potential. The increasing of the anodic overpotential window of the composite film makes possible to use these films as catalytic substrates to carry on reactions, like ethanol or methanol oxidation, without significant film deterioration.

The inorganic particles significantly influence both the oxidation stability and morphology of the PPy matrix. Incorporation of platinum and magnetite nanoparticles into the polypyrrole film increase the current density of the electrochemical response and shift the oxidation potential towards more anodic values. Moreover, the deposition of platinum and magnetite allows larger over-potential window to promote redox reactions without compromising the stability to oxidation of the polymer.

7. Acknowledgments

This work is partially supported by the University of Antioquia. The financial support from the Excellence Center for Novel Materials –CENM- and from COLCIENCIAS is also greatly appreciated by the authors.

8. References

Andrieux, C. P., Audebert, P., Hapiot, P., and Saveant, J. M. (1991). Identification of the first steps of the electrochemical polymerization of pyrroles by means of fast potential step techniques. *The Journal of Physical Chemistry* 95, 10158-10164.

Benner, L. S., Suzuki, T., Meguro, K., and Tanaka, S. (1991). Precious metals: science and technology. *International Precious Metals Institute(United States), 1991*, 799.

Bouzek, K., Mangold, K. M., and Jüttner, K. (2001a). Electrocatalytic activity of platinum modified polypyrrole films for the methanol oxidation reaction. *Journal of applied electrochemistry* 31, 501-507.

Bouzek, K., Mangold, K. M., and Jüttner, K. (2001b). Platinum distribution and electrocatalytic properties of modified polypyrrole films. *Electrochimica acta* 46, 661-670.

Bravo-Grimaldo, E., Hachey, S., Cameron, C. G., and Freund, M. S. (2007). Metastable reaction mixtures for the in situ polymerization of conducting polymers. *Macromolecules* 40, 7166-7170.

Chen, A. H., Wang, H. Q., and Li, X. Y. (2004). Influence of concentration of FeCl3 solution on properties of polypyrrole-Fe3O4 composites prepared by common ion absorption effect. *Synthetic Metals* 145, 153-157.

Chen, A. H., Wang, H. Q., Zhao, B., and Li, X. Y. (2003). Preparation of polypyrrole-Fe3O4 nanocomposites by the use of common ion effect. *Synthetic Metals* 139, 411-415.

Demets, G. J. F., Anaissi, F. J., and Toma, H. E. (2000). Electrochemical properties of assembled polypyrrole/V2O5 xerogel films. *Electrochimica acta* 46, 547-554.

Deng, J., Peng, Y., He, C., Long, X., Li, P., and Chan, A. S. C. (2003a). Magnetic and conducting Fe3O4–polypyrrole nanoparticles with core shell structure. *Polymer international* 52, 1182-1187.

Deng, J. G., He, C. L., Long, X. P., Peng, Y. X., Li, P., and Chan, A. S. C. (2003b). Preparation and characterization of magnetic Fe_3O_4- polypyrrole nanoparticles. *Acta Polymerica Sinica*, 393-397.

Deng, J. G., Peng, Y. X., He, C. L., Long, X. P., Li, P., and Chan, A. S. C. (2003c). Magnetic and conducting Fe3O4-polypyrrole nanoparticles with core-shell structure. *Polymer International* 52, 1182-1187.

Elliott, C. M., Kopelove, A. B., Albery, W. J., and Chen, Z. (1991). Nonaqueous electrochemistry of polypyrrole/polystyrenesulfonate composite films: voltammetric, coulometric, EPR, and ac impedance studies. *The Journal of Physical Chemistry* 95, 1743-1747.

Ferreira, C., Domenech, S., and Lacaze, P. (2001). Synthesis and characterization of polypyrrole/TiO2 composites on mild steel. *Journal of applied electrochemistry* 31, 49-56.

Furukawa, Y., Tazawa, S., Fujii, Y., and Harada, I. (1988). Raman spectra of polypyrrole and its 2, 5-13C-substituted and C-deuterated analogues in doped and undoped states. *Synthetic metals* 24, 329-341.

Garcia, B., Lamzoudi, A., Pillier, F., Le, H. N. T., and Deslouis, C. (2002). Oxide/polypyrrole composite films for corrosion protection of iron. *Journal of the Electrochemical Society* 149, B560-B566.

Georgieva, M., and Andonovski, B. (2003). Determination of platinum (IV) by UV spectrophotometry. *Analytical and bioanalytical chemistry* 375, 836-839.

Giacomelli, C., Giacomelli, F., Baptista, J., and Spinelli, A. (2004). The effect of oxalic acid on the corrosion of carbon steel. *Anti-Corrosion Methods and Materials* 51, 105-111.

Holzhauser, P., Bouzek, K., and Bastl, Z. (2005). Electrocatalytic properties of polypyrrole films prepared with platinate (II) counter-ions. *Synthetic metals* 155, 501-508.

Huang, Z., Tang, F., and Zhang, L. (2005). Morphology control and texture of Fe3O4 nanoparticle-coated polystyrene microspheres by ethylene glycol in forced hydrolysis reaction. *Thin Solid Films* 471, 105-112.

Iroh, J. O., and Su, W. (2000). Corrosion performance of polypyrrole coating applied to low carbon steel by an electrochemical process. *Electrochimica acta* 46, 15-24.

Iroh, J. O., and Su, W. (2002). Adhesion of electrochemically formed polypyrrole coatings to low carbon steel. *Journal of applied polymer science* 85, 2757-2763.

Kawai, K., Mihara, N., Kuwabata, S., and Yoneyama, H. (1990). Electrochemical Synthesis of Polypyrrole Films Containing TiO Powder Particles. *Journal of the Electrochemical Society* 137, 1793.

Kuwabata, S., Masui, S., Tomiyori, H., and Yoneyama, H. (2000). Charge-discharge properties of chemically prepared composites of V2O5 and polypyrrole as positive electrode materials in rechargeable Li batteries. *Electrochimica acta* 46, 91-97.

Lamprakopoulos, S., Yfantis, D., Yfantis, A., Schmeisser, D., Anastassopoulou, J., and Theophanides, T. (2004). An FTIR Study of the Role of H2O and D2O in the Aging Mechanism of Conductive Polypyrroles. *Synthetic metals* 144, 229-234.

Lehr, I., and Saidman, S. (2007). Corrosion protection of iron by polypyrrole coatings electrosynthesised from a surfactant solution. *Corrosion science* 49, 2210-2225.

Lenz, D. M., Delamar, M., and Ferreira, C. A. (2003). Application of polypyrrole/TiO2 composite films as corrosion protection of mild steel. *Journal of Electroanalytical Chemistry* 540, 35-44.

Lu, X., Chao, D., Chen, J., Zhang, W., and Wei, Y. (2006). Preparation and characterization of inorganic/organic hybrid nanocomposites based on Au nanoparticles and polypyrrole. *Materials letters* 60, 2851-2854.

Martins, N., Moura e Silva, T., Montemor, M., Fernandes, J., and Ferreira, M. (2008). Electrodeposition and characterization of polypyrrole films on aluminium alloy 6061-T6. *Electrochimica Acta* 53, 4754-4763.

Marín, T., Calderón, J. A., and Jaramillo, F. (2010). Study of polypyrrole/platinum composite films. *Rev. Fac. Ing. Univ. Antioquia*, 49-56.

Marín, T., Montoya, P., Jaramillo, F., and Calderón, J. (2009). Evaluación de la incorporación de partículas de magnetita y platino en la estabilidad de películas de polipirrol. *Revista Latinoamericana de Metalurgia y Materiales* S1.

McNally, E., Zhitomirsky, I., and Wilkinson, D. (2005). Cathodic electrodeposition of cobalt oxide films using polyelectrolytes. *Materials chemistry and physics* 91, 391-398.

Montoya, P., Jaramillo, F., Calderón, J., Córdoba de Torresi, S., and Torresi, R. (2010). Evidence of redox interactions between polypyrrole and Fe3O4 in polypyrrole-Fe3O4 composite films. *Electrochimica Acta* 55, 6116-6122.

Montoya, P., Jaramillo, F., Calderón, J., de Torresi, S. I. C., and Torresi, R. (2009). Effect of the Incorporation of Magnetite Particles in Polypyrrol Films. *Portugaliae Electrochimica Acta* 27, 337-344.

Nie, J., Tallman, D. E., and Bierwagen, G. P. (2008). The electrodeposition of polypyrrole on Al alloy from room temperature ionic liquids. *Journal of Coatings Technology and Research* 5, 327-334.

Ocon, P., Cristobal, A., Herrasti, P., and Fatas, E. (2005). Corrosion performance of conducting polymer coatings applied on mild steel. *Corrosion science* 47, 649-662.

Okner, R., Domb, A. J., and Mandler, D. (2007). Electrochemical formation and characterization of copolymers based on N-pyrrole derivatives. *Biomacromolecules* 8, 2928-2935.

Pailleret, A., Hien, N. T. L., Thanh, D. T. M., and Deslouis, C. (2007). Surface reactivity of polypyrrole/iron-oxide nanoparticles: electrochemical and CS-AFM investigations. *Journal of Solid State Electrochemistry* 11, 1013-1021.

Radhakrishnan, S., and Adhikari, A. (2006). Role of dopant ions in electrocatalytic oxidation of methanol using conducting polypyrrole electrodes. *Journal of power sources* 155, 157-160.

Santos, M., Brolo, A., and Girotto, E. (2007). Study of polaron and bipolaron states in polypyrrole by in situ Raman spectroelectrochemistry. *Electrochimica acta* 52, 6141-6145.

Su, W., and Iroh, J. O. (1997). Kinetics and efficiency of aqueous electropolymerization of pyrrole onto low carbon steel. *Journal of Applied Polymer Science* 65, 617-624.

Tamm, J., Johanson, U., Marandi, M., Tamm, T., and Tamm, L. (2004). Study of the properties of electrodeposited polypyrrole films. *Russian Journal of Electrochemistry* 40, 344-348.

Vishnuvardhan, T., Kulkarni, V., Basavaraja, C., and Raghavendra, S. (2006). Synthesis, characterization and ac conductivity of polypyrrole/Y 2 O 3 composites. *Bulletin of Materials Science* 29, 77-83.

Wallace, G. G. (2003). "Conductive electroactive polymers: intelligent materials systems," CRC.

Wencheng, S., and Iroh, J. O. (1998). Effects of electrochemical process parameters on the synthesis and properties of polypyrrole coatings on steel. *Synthetic metals* 95, 159-167.

Wu, T. M., Yen, S. J., Chen, E. C., Sung, T. W., and Chiang, R. K. (2007). Conducting and magnetic behaviors of monodispersed iron oxide/polypyrrole nanocomposites synthesized by in situ chemical oxidative polymerization. *Journal of Polymer Science Part a-Polymer Chemistry* 45, 4647-4655.

Yakushi, K., Lauchlan, L., Clarke, T., and Street, G. (1983). Optical study of polypyrrole perchlorate. *The Journal of chemical physics* 79, 4774.

8

Conducting Polypyrrole Shell as a Promising Covering for Magnetic Nanoparticles

Alexandrina Nan, Izabell Craciunescu and Rodica Turcu
National Institute of Research and Development for Isotopic and Molecular Technologies
Romania

1. Introduction

Conducting polymers are a recent generation of polymers, opening the progress in understanding the fundamental chemistry and physics of π-conjugated macromolecules. Among conducting polymers, polypyrrole (PPy) has attracted great interest owing to its high conductivity and relatively high environmental stability, therefore the potential applications of polypyrrole are numerous. The combination of PPy with other inorganic materials in order to prepare composites which combine the properties of both materials is a very promising way to extend the application field of PPy but also of the inorganic materials. In order to improve the poor processability of polypyrrole a lot of methods have been explored in the preparation of soluble or swollen PPy (Masuda et al., 1989; Stanke et al., 1993) and dispersible fine powdered PPy (Bjorklund & Liedberg, 1986; Armes, et al., 1987; Cawdery et al., 1988; Aldissi & Armes, 1991). To extend the application field of polypyrrole, sterically stabilized PPy colloids were synthesized in aqueous media by chemically polymerizing pyrrole monomers in the presence of a suitable water-soluble polymer, such as methyl cellulose or poly(vinyl alcohol) (Armes et al., 1987; Armes & Vincent, 1987). The preparation of core-shell structures based on polypyrrole were first reported by Yassar et al. long before polypyrrole based magnetic core shell nanoparticles were developed. The authors coated latex particles by conducting polypyrrole (Yassar et al., 1987). After this many research groups reported the preparation of colloidal conducting polypyrrole by coating particles with a thin layer of conjugated polypyrrole to form conducting composites with a core-shell structure. Most of them were non magnetic. Different types of polymers, metals and metal oxides were used as shells. Many papers describe the polymerization of pyrrole in the presence of different polymers structure, polystyrene being the most often used. (Lascelles & Armes, 1995; Lascelles et al., 1997; Lascelles & Armes, 1997; Cairns & Armes 1999; Lu et al., 2003; Bousalem et al., 2003; Bousalem et al., 2004a; Bousalem et al., 2004b; Bousalem et al., 2005; Benabderrahmane et al., 2005; Mangeney et al., 2006; Yip et al., 2006; Lee et al., 2009; Wang et al., 2009).
Latex particles having a poly(butyl methacrylate) (PBMA) core of about 700 nm and a very thin polypyrrole (PPy) shell were reported by Huijs & Lang 2000. Further studies illustrated the effect of the thickness of the polypyrrole shell on latex properties (Huijs et al., 2001). Other ways to water-based processable conducting polypyrrole are based on the preparation of colloidal core-shell polypyrrole nanoparticles by oxidative polymerization of pyrrole in the presence of ultrafine silica nanoparticles. (Maeda & Armes, 1994; Azioune et

al., 1999; Yang et al., 2006; Marini et al., 2008; Liu et al., 2008; Pourabbasa et al., 2010) or the copolymerization of pyrrole and functionalized pyrroles in the presence of the such silica nanoparticles (McCarthy et al., 1997; Goller et al., 1998; Azioune et al., 2004).

In 1995 Maeda & Armes presented a study about the chemical polymerization of pyrrole in the presence of various other ultrafine inorganic oxide sols such as tin(IV) oxide, zirconium, antimony(V) oxide, yttrium, and titanium(IV) oxide. It turned out that only the tin(IV) oxide sols act as effective particulate dispersants; the other four oxide systems failed to prevent macroscopic precipitation. SiO_2@CdSe/polypyrrole multi-composite core-shell structures were synthesized by cationic polymerisation of pyrrole (Hao et al., 2006).

A simple non-template one-step method for the synthesis of 2.0-2.5 nm palladium nanoparticles encapsulated into a polypyrrole shell is based on direct redox reaction between palladium(II) acetate and pyrrole in acetonitrile medium (Vasilyeva et al., 2008). They stipulated that palladium nanoparticles are found to be able to self-organize into spherical Pd/PPy composites.

Selenium-polypyrrole core-shell nanostructures fabricated by *in-situ* polymerization process and functionalized with trasferrin can be used for targeting and imaging of human cervical cancer cells (Li & Liu, 2008). Silver is another type of metal which was used as a core for polypyrrole shells (Feng et al., 2007; Wang & Shi 2007; Rojas et al., 2008; Ye & Lu et al., 2008; Ye et al., 2009). Coating gold nanoparticles with conductive polypyrrole resulted in uniform core-shell nanoparticles with tailored core aggregation and shell thickness (Xing et al., 2009).

An extremely important class of core-shell nanoparticles based on polypyrrole are magnetic nanoparticles. Although such composites share many aspects with non-magnetic nanoparticles they exhibit magnetism a unique property which is extremely important for many applications. In the present review the synthesis, characterization and applications of this class of magnetic core shell nanostructure are illustrated.

2. Preparation of magnetic core-shell nanostructure based on conducting polypyrrole

2.1 Preparation of magnetic core-shell nanostructured based on unfunctionalized polypyrrole

First publication which reported about the preparation of magnetic core-shell nanoparticles based on conducting polypyrrole appeared in 1994. Later on a promising approach using the direct polymerization of pyrrole in the presence of Fe_3O_4 nanoparticles as reported by Deng et al. (Scheme 1) (Deng et al., 2003). The magnetite was obtained by partial oxidation of iron(II) followed by the polymerisation of the pyrrole using iron(III) chloride as oxidant. In this way novel nanocomposites with a well-defined core-shell microstructure were obtained

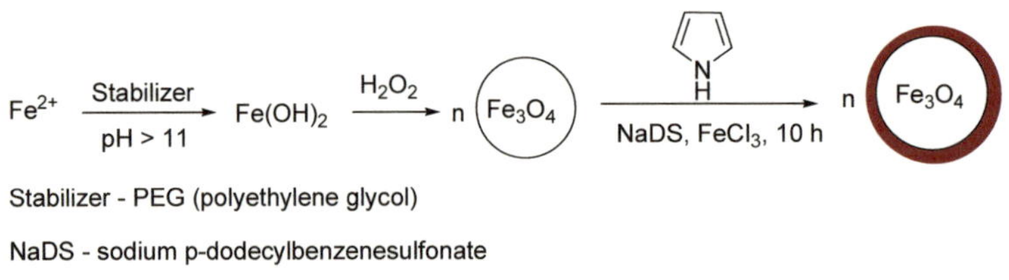

Scheme 1. Preparation of magnetic-conducting polypyrrole core-shell structure

Core-shell nanostructures based on the same polypyrrole and spherical hydroxyl iron (Fe[OH]) cores were prepared by Li et al. through the polymerization of pyrrole in the presence of *p*-toluenesulfonic acid (*p*-TSA) as dopant and Fe[OH] with a diameter 0.5-5 μm (Li et al., 2006). They obtained a micro/nanostructure which exhibit high conductivity ($\sim\sigma_{max}$ = 50.6 S/cm) and superparamagnetic behaviour.
Ultrasonic irradiation technique was applied to polypyrrole/Fe_3O_4 nanocomposite preparation, for a better dispersion of the Fe_3O_4 nanoparticles resulting in nanocomposite with high conductivity. Again, the polymerization of pyrrole was mediated by $FeCl_3$ as oxidant (Qui et al., 2006).
An efficient DNA retrieval system was developed based on Fe_3O_4 polypyrrole nanoparticles. Here, the surface of the magnetite was modified with pyrrole-2-carboxylic acid followed by free radical polymerization of unsubstituted pyrrole., The resulting Fe_3O_4@polypyrrole showed excellent high affinity to DNA (Park et al., 2008). Superparamagnetic Fe_3O_4 nanoparticles were successfully encapsulated inside polypyrrole via an emulsion polymerization using polyvinyl alcohol as a surfactant, yielding magnetic core-shell nanostructurs of spherical shape with a diameter of 80-100 nm. These products exhibited high magnetization values and good electrical conductivities (Wuang et al., 2007). Cell targeting nanostructures were obtained when the surfaces of these nanoparticles were further functionalized with folic acid. Their potential for targeting of cancer cells was investigated revealing that the uptake by MCF-7 breast cancer cells is significantly enhanced as compared to non-functionalized precursors (Wuang et al., 2007).
Uniform Fe_3O_4 nanospheres with a diameter of 100 nm could be obtained by the microwave solvothermal method. When these nanospheres were covered with a polypyrrole shell followed by the electrostatic interactions with citrate stabilized gold nanoparticles (Au NP) a type of three-component Fe_3O_4/polypyrrole/Au nanocomposites with core/shell/shell structure were obtained (Zhang et al., 2008).
Another important application of the core-shell magnetic nanoparticles based on magnetite and unfunctionalied polypyrrole is the removal of fluoride ions from aqueous solution. The magnetic property makes the material easy to retrieve from solution using external magnetic field (Bhaumik et al., 2011).
As shown by our group polypyrrole magnetite core shell nanoparticles can efficiently be synthesized if pyrrole is polymerized by ammonium persulfate (APS) as oxidant in water based magnetic nanofluids (MF) (Scheme 2) (Turcu et. al., 2006).

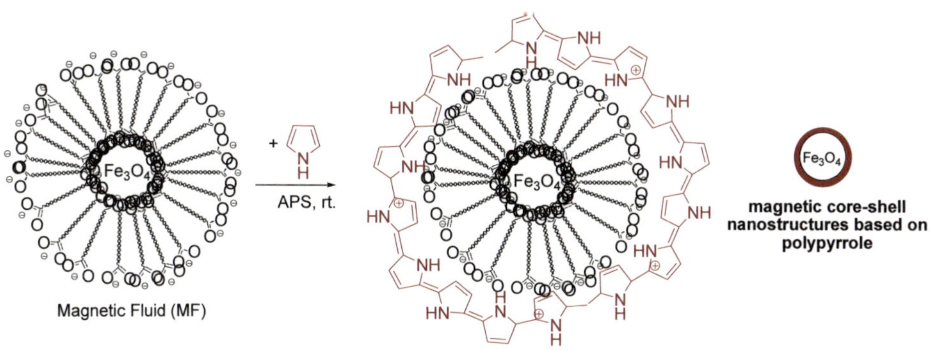

Scheme 2. Preparation of magnetic core-shell nanostructure based on polypyrrole and magnetic fluid

These nanofluids used in the polymerization reaction, consist of magnetite nanoparticles which are covered and thus stabilized by fatty acid shells rather than of "naked" nanoparticles. The preparation of the magnetic fluid starts with coprecipitation of Fe^{3+}, Fe^{2+} ions in NH_4OH-solution to give magnetite nanoparticles. The temperature was maintained at 80–82 ^0C, in order to obtain entirely magnetite nanoparticles and to optimize the following chemisorption of the surfactant (Bica et al., 2007). Combinations of surfactants with different chain lengths: myristic acid (MA), lauric acid (LA) and dodecylbenzenesulphonic acid (DBS) were used, such as MA+MA, LA+LA, MA+DBS, DBS+DBS, LA+DBS for the preparation of polypyrrole based nanocomposites (Turcu et al., 2008). The electrical conductivity of the nanocomposites can be controlled by the ratio of the starting materials.

Tubular Fe_3O_4/polypyrrole nanocomposites can be synthesized by in situ polymerization of pyrrole in the presence of the monodispersed 4 nm Fe_3O_4 nanoparticles (Wu et al., 2007). The monodispersed 4 nm Fe_3O_4 nanoparticles which served as cores were synthesized using the thermal decomposition of a mixture of iron (III) acetylacetonate and oleic acid in the presence of high boiling solvents. The resulting nanoparticles were further dispersed in an aqueous solution with the anionic surfactant sodium bis(2-ethylhexyl)sulfosuccinate to form micelle/Fe_3O_4 spherical templates that avoid the aggregation of Fe_3O_4 nanoparticles during the further preparation of the nanocomposites. The Fe_3O_4/PPy nanocomposites were then synthesized via in situ chemical oxidative polymerization on the surface of the spherical templates.

A new type of magnetic core-shell nanostructures based on polypyrrole was developed by using iron–gold (Fe@Au) nanoparticles instead of magnetite as magnetic cores. The gold-coated iron nanoparticles generally were obtained by reverse micelle method, wherein cetyltrimethylammonium bromide (CTAB) acted as surfactant and 1-butanol as co-surfactant (Pana et al., 2007). The resulting Fe@Au nanoparticles were covered by the surfactant and were further coated with polypyrrole by APS-mediated polymerization of pyrrole.

Another type of core-shell nanoparticles were obtained with flaky $BaFe_{12}O_{19}$ nanoparticles (10-20 nm in thickness) as polymerization seeds, functionalized and microstructured quasi-spherical $BaFe_{12}O_{19}$/polypyrrole organic-inorganic composites were prepared by a conventional in situ chemical oxidative polymerization (Xu et al., 2008; Birsöz et al., 2010). Magnetic core-shell nanoparticles based on polypyrrole and $SrFe_{12}O_{19}$ nanoparticles composites were prepared by *in situ* polymerization method; the morphology of such magnetic nanoparticles could be modified from sphere-like to conglobulation-like and arborisation-like structure increasing the pyrrole/$SrFe_{12}O_{19}$ mass ratio (Zhang et al., 2009).

A polypyrrole/ferrospinel ($NiFe_2O_4$) nanocomposite with a core-shell structure was prepared by the in situ chemical polymerization of pyrrole in the presence of $NiFe_2O_4$ nanoparticles in water-in-oil microemulsion (Jiang et al., 2010).

$ZnFe_2O_4$/polypyrrole core-shell nanoparticles could be subsequently synthesized via in situ chemical oxidative polymerization of pyrrole monomers on the surface of $ZnFe_2O_4$ nanoparticles using ammonium persulfate as oxidant at relatively low temperature 80 ^0C. The shell thickness of core-shell nanoparticles could be easily controlled by adjusting the amount of pyrrole monomer. The synthesis was shown to be inexpensive, nontoxic and reproducible solvothermal method (Li et al., 2009).

2.2 Preparation of magnetic core-shell nanostructured based on functionalized polypyrrole

By using functionalized pyrrole instead of unsubstituted pyrrole the applicability of magnetic core shell nanoparticles can be extended considerably. Corresponding pyrrole monomers are available either commercially or by various chemical reactions.

The first publication about the preparation and characterization of core-shell magnetic nanostructured appeared in 1994 when Nguyen and Diaz reported a simple route to poly(pyrrole-*N*-propylsulfonate) polymer containing nano-sized γ-Fe_2O_3 particles (Nguyen & Diaz, 1994). The preparation occurred in two steps by first polymerization of sodium pyrrole-*N*-propylsulfonate using excess of $FeCl_3$ as oxidant resulting a black powder poly(pyrrole-*N*-propylsulfonate). In the second step, the magnetic-polypyrrole nanostructures were formed by adding an ammonia solution to the preformed polymer and keeping the reaction under stirring and at 70 ^{0}C for 30 minutes while the γ-Fe_2O_3 is formed (Scheme 3).

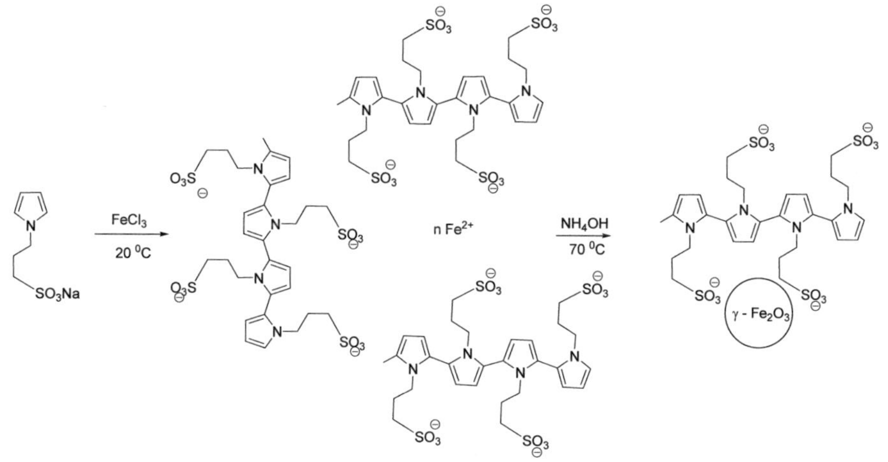

Scheme 3. Preparation of magnetic nanoparticles based on γ-Fe_2O_3 and poly(pyrrole-*N*-propylsulfonate)

Although Fe_2O_3–poly(pyrrole-*N*-propylsulfonate) nanostructures have been successfully prepared they still have low room-temperature conductivity (10^{-4}–10^{-1} Scm^{-1}) and low coercive force. In addition, their structures and properties are difficult to control owing to the complicated synthetic method involved, therefore further development of synthetic methods to produce novel electrical-magnetic nanostructures with high room-temperature conductivity and coercive force was highly required.

External magnetic field was applied to affect assembling of functionalized polypyrrole-coated latex particles. Here *γ*-Fe_2O_3 superparamagnetic nanoparticle-containing polystyrene cores were coated by a functionalized polypyrrole shell. The polystyrene particles provide a rigid, non-deformable and nonporous support for the polypyrrole coating (Mageney et al., 2007).

Thus Nan et al. synthesized pyrroles which are linked to amino acids at the β-position of the pyrrole ring (Scheme 4) (Nan et al., 2008).

These amino acid pyrrole conjugates were successfully used in the synthesis of novel functionalized magnetic nanostructures by chemical oxidative polymerization in aqueous solutions in the presence of Fe_3O_4 nanofluids (Scheme 5) (Nan et al., 2008).

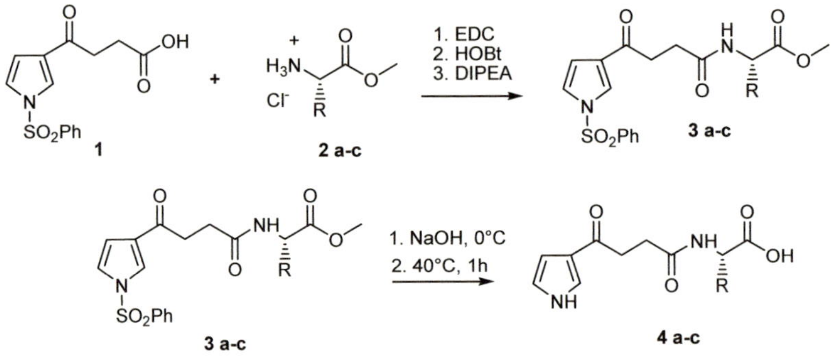

2a, 3a, 4a - R = *sec*-Bu
2b, 3b, 4b - R = *i*-Pr
2c, 3c, 4c - R = CH_2-Ph

Scheme 4. Synthesis of new pyrrole monomers functionalized by α-amino acids

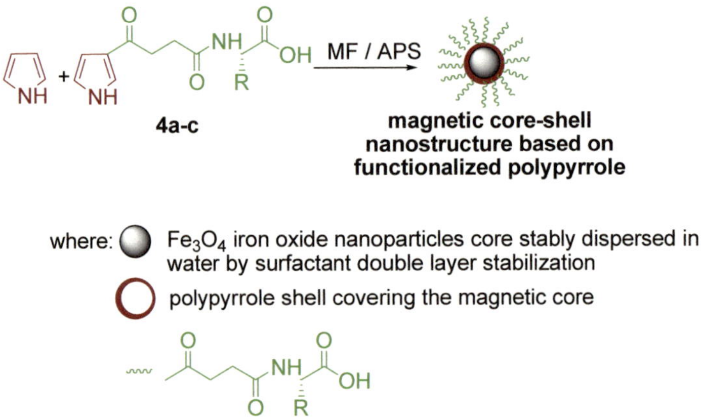

Scheme 5. Preparation of magnetic core-shell nanostructures functionalized by α-amino acids

Glycyl-leucine dipeptide and bovine serum albumin were also used for the functionalization of the polypyrrole shell which covered the magnetic nanoparticles (Nan et al., 2010) using N-hydroxysuccinimide as activating reagent. In the case of the dipeptide, this were first covalent attached on the pyrrole ring and after that the polymerization of functionalized pyrrole monomer in the presence of a ferrofluid based on water were performed (Scheme 6).

The BSA-functionalized magnetic core shell nanoparticles were synthesized starting from the polymerization of 3-(Pyrrol-1-yl)-propanoyl-*N*-hydroxysuccinate in the presence of double layer stabilized magnetic ferrofluid in water, issuing a *N*-hydroxysuccinate surface active magnetic nanoparticles. The resulted magnetic nanoparticle based on functionalized polypyrrole with *N*-hydroxysuccinate was dispersed in phosphate-buffered saline solution (PBS, pH 7.4) and then, BSA was added. In this way BSA substituted a part of the *N*-hydroxysuccinate groups while the others survived and gave propionic acid moieties by hydrolysis.

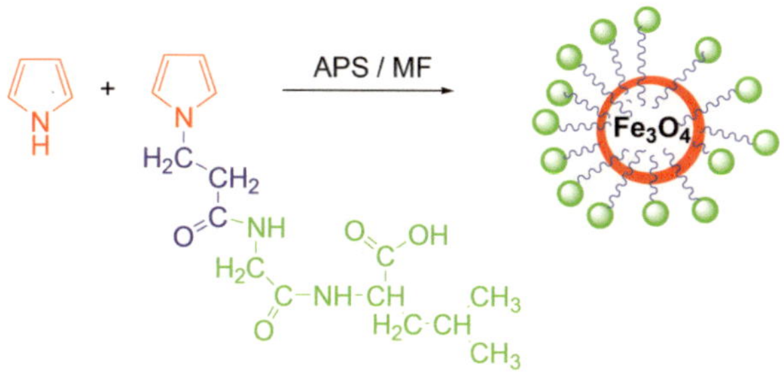

MF - magnetic nanofluid water based

APS - ammonium persulfate

Scheme 6. The functionalization of magnetic nanoparticles based on functionalized polypyrrole with glycyl-leucine dipeptide

For instance, the popular *N*-hydroxysuccinimide ester is prone to hydrolysis before and during the coupling reaction, which can both reduce coupling yields and make the yields and the material irreproducible. Therefore, it is highly desirable and useful to develop new reactions which can be easily perform under mild condition, with high yields and without by products. These requirements are met by the Medal-Sharpless cycloadition. This 1,3-dipolar cycloaddition reaction is recognized as the best example of click chemistry and can be performed to give a quantitative yield, in multiple solvents (including water) and in the presence of various functional groups, as well as under mild reaction conditions. As compared with other tethering tools, the Meldal-Sharpless cycloaddition method tolerates many functional groups allowing the omission of protective groups and thus was also widely applied to biological molecules.

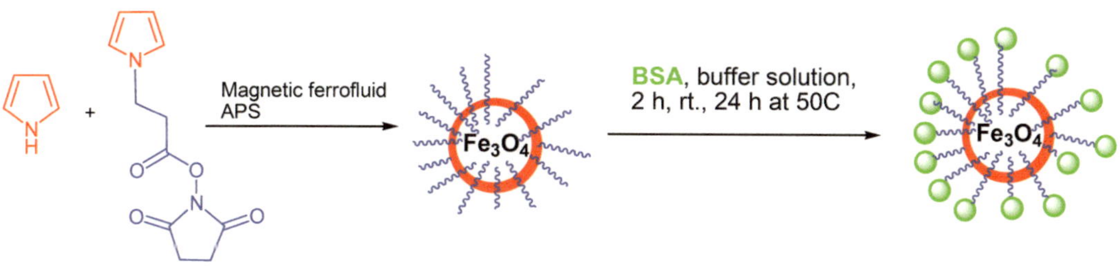

Scheme 7. Immobilization of albumin on the magnetic core-shell nanoparticles based on polypyrrole

The mild and versatile method based on Cu-catalyzed [3+2]-cycloaddition was developed to tether biomolecules, such as monosaccharides, biotin, cholesterol or uridine to the *N*-atom of pyrrole. The required azido and alkyne function can be placed in either reactant, i. e. in the pyrrole or the biomolecule. The resulting products were employed as precursors for functionalized superparamagnetic polypyrrole core-shell nanoparticles (Karsten et al., 2010). The functionalizations of the core-shell nanoparticles were prepared in two ways. First, the biomolecules are attached on the pyrrole monomer via "click-reaction" and after this take place the polymerization of the preformed pyrrole monomer in the presence of the magnetic fluid (scheme 8).

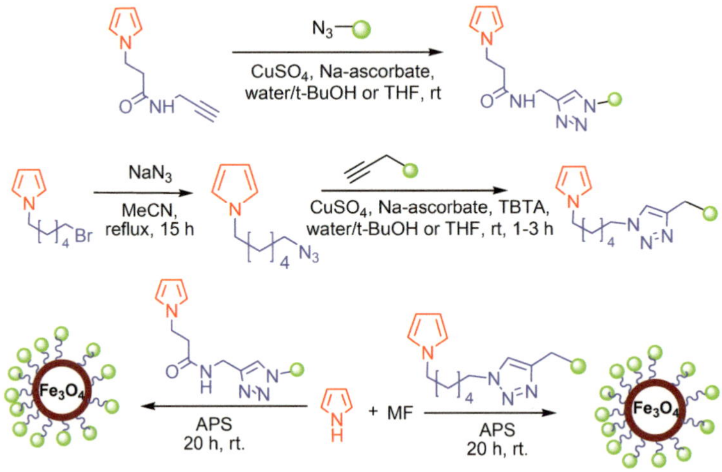

Scheme 8. Preparation of magnetic core-shell nanoparticles based on biofunctionalized polypyrrole

Another way to obtained magnetic core-shell nanoparticles based on functionalized polypyrrole using "click-chemistry"occurs in two steps. In the first step take place the polymerization of the pyrrole monomer which has attached on the ring azide group or alchine group in the presence of magnetic nanofluid. The second strep consist in the execution of the Cu-catalyzed [3+2]-cycloaddition with different biomolecules on the beforehand prepared magnetic core-shell nanoparticles based on functionalized polypyrrole with azide or alchine groups (scheme 9).

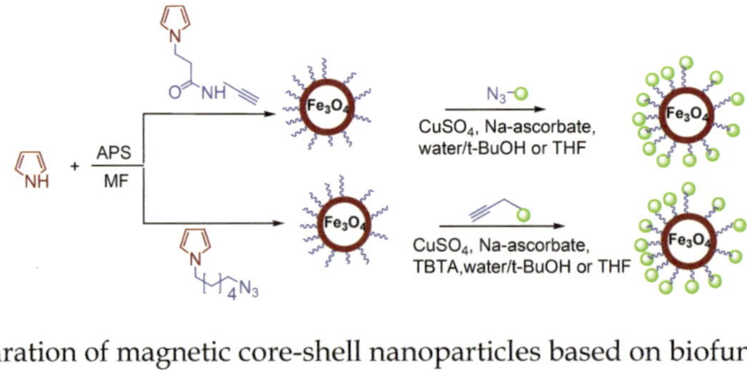

Scheme 9. Preparation of magnetic core-shell nanoparticles based on biofunctionalized polypyrrole

As was mentioned before using Cu-catalyzed [3+2]-cycloaddition reaction, many types of biomolecules with interesting application can be attached on different substrates; this principal was applied also for the magnetic core-shell nanostructured based on polypyrrole. In the scheme 10 are showed which types of biomolecules where attached on the magnetic core-shell nanoparticles (Karsten et al., 2010).

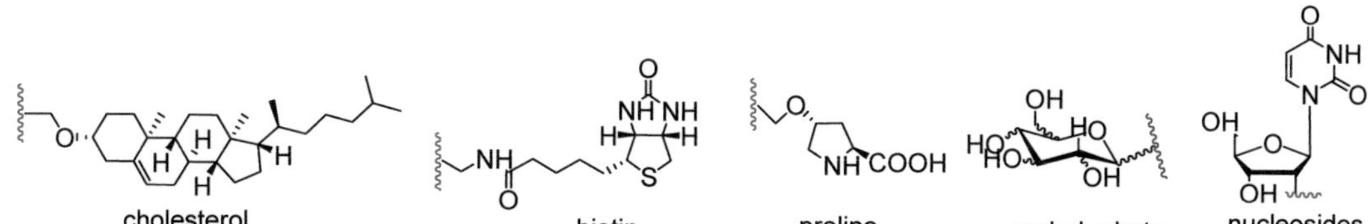

Scheme 10. Biomolecules which were attached on the magnetic core-shell nanostructures based on polypyrrole using the „click-chemistry"

3. Methods for characterization of magnetic core-shell nanoparticles based on polypyrrole

For magnetic core-shell nanoparticles based on polypyrrole the characterization methods which can be applied for a better structural investigation are: morphological investigation, FTIR spectroscopy, X-ray diffraction, X-photoelectron spectroscopy and the magnetic properties.

3.1 Morphologic characterization of magnetic core-shell nanoparticles based on polypyrrole

The size and morphology of the magnetic core-shell nanoparticles were mainly characterized by Transmission Electron Microscopy (TEM), High-Resolution Transmission Electron Microscopy (HRTEM), field-emission scanning electron microscopy (FESEM), but in some cases is also used Scanning Electron Microscopy (SEM).

The SEM microscopy gives more an overview of the size and shape of the magnetic core-shell nanoparticles based on polypyrrole. A very clear SEM image (figure 1) in which the coating phenomenon can be clearly observed is presented in the case of $ZnFe_2O_4$ cores surrounded by polypyrrole shells. In the SEM images can be see that the $ZnFe_2O_4$ core diameter is around 80 nm and after polymerization of pyrrole monomer the diameter increased at about 100–300 nm (Li et al., 2009).

Fig. 1. The SEM images of (a) $ZnFe_2O_4$ nanoparticles and (b) magnetic core-shell nanoparticles based on $ZnFe_2O_4$ andpolypyrrole by using 1 ml of pyrrole at 80 ^{0}C for 8 h

The TEM microscopy provides more clear information about the size and shape than SEM microscopy. In the literature TEM microscopy is a useful tool to determine the magnetic core diameter and the thickness of polypyrrole shell, but in the same time TEM allows the comparison between the former magnetic nanoparticles and magnetic core-shell nanoparticles based on polypyrrole. Figure 2 shows the difference between the TEM image of the ferrofluid used for the polymerization of unfunctionalized pyrrole monomer and the TEM image (figure 2a) of the magnetic core-shell nanoparticles based on polypyrrole and magnetite (figure 2b).

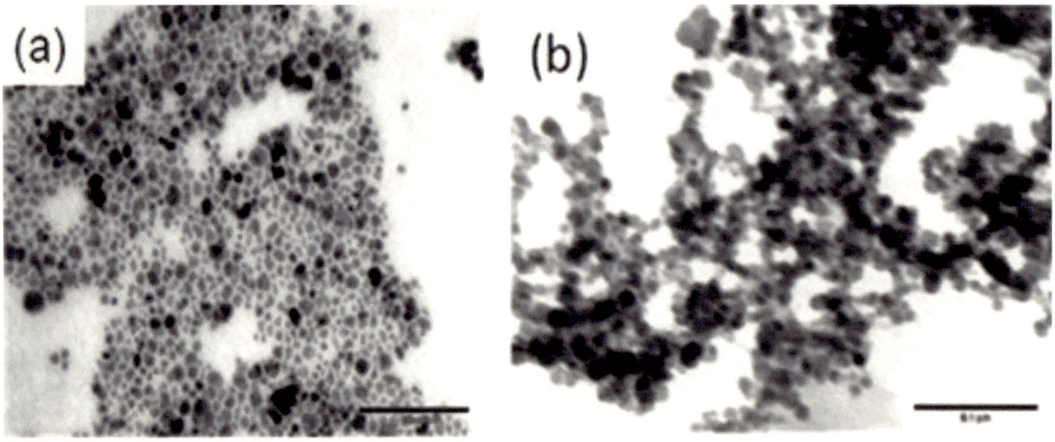

Fig. 2. The TEM images of (a) magnetic ferrofluid and (b) magnetic core-shell nanoparticles prepared based on the ferrofluid and unfunctionalized polypyrrole

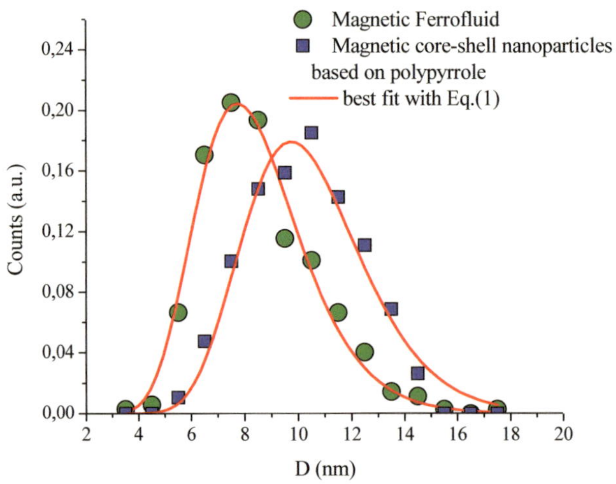

Fig. 3. The corresponding distributions of diameters for magnetic ferrofluid and of the magnetic core-shell nanoparticles based on polypyrrole

The analysis of the TEM images presented above enabled the determination of the diameters distributions of the nanoparticles, as shown in the figure 3. The normalized distribution of diameters is well described by a lognormal distribution function (equation 1):

$$f(D)=\frac{1}{D\sigma\sqrt{2\pi}}\exp\left[-\frac{1}{2}\left(\frac{\ln^2 D/D_0}{\sigma^2}\right)\right] \quad (1)$$

where D is the diameter, D_0 is the mean diameter and σ is the standard deviation.
From the diameter histogram were obtained a mean diameter D_0 = 8.2 nm for the magnetic ferrofluid and a mean diameter D_0 = 10.2 nm for the magnetic core-shell nanoparticles based on polypyrrole, that means that the thickness of the polypyrrole shell it is about 2 nm (Turcu et al., 2008).
Nevertheless HRTEM with it high resolution makes it ideal for imaging materials on the atomic scale. The HRTEM images given in Figure 4 presents the core-shell nanoparticles based on unfunctionalized polypyrrole (Figure 4a), with a thin shell around 2 - 3 nm, which is similar to those found in the magnetic core-shell nanoparticles formed with substituted pyrrole (Figure 4b) (Nan et al., 2008).

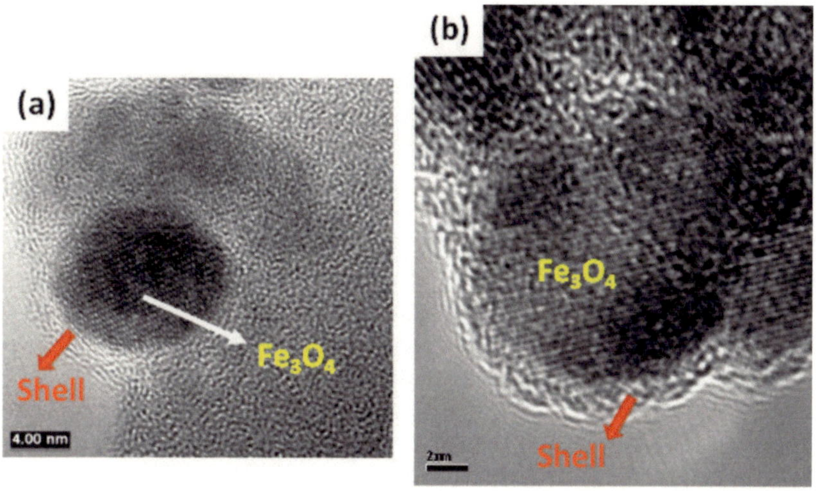

Fig. 4. HRTEM images of (a) magnetic core-shell nanoparticles based on unfunctionalized polypyrrole and (b) magnetic core-shell nanoparticles based on functionalized polypyrrole

The HRTEM microscopy point out also that the polymer which surrounds the magnetite nanoparticles seems to be strongly adhesive onto the surface of the nanoparticles resulting in a very intimate connection between the two components.

3.2 X-ray diffraction of the magnetic core-shell nanoparticles based on polypyrrole

An X-ray diffraction method is capable to determine average particle size, microstrains, probability of faults as well as the particle size distribution of magnetic nanoparticles. By XRD method one can be obtain the crystallite size that has different values for the different crystallographic planes. There is a large difference between the grain size and crystallite size due to the physical meaning of the two concepts. Practically speaking, it is not easy to obtain accurate values of the crystallite size and microstrain without extreme care in experimental measurements and analysis of XRD data. In the figure 5 presented the XRD of two samples of magnetic core-shell nanoparticles based on polypyrrole prepared by the oxidative polymerization of pyrrole monomer in the presence of magnetic ferrofluid (Aldea et al., 2009). The characteristic peaks for Fe_3O_4 can be clearly observed in the XRD spectra from figure 5. In addition, the characteristic peak (200) for FeO appears. Moreover, the intense peak at 2θ = 35.4 degrees could be due to the superposition of Fe_3O_4 characteristic peak (311) and FeO characteristic peak (111).

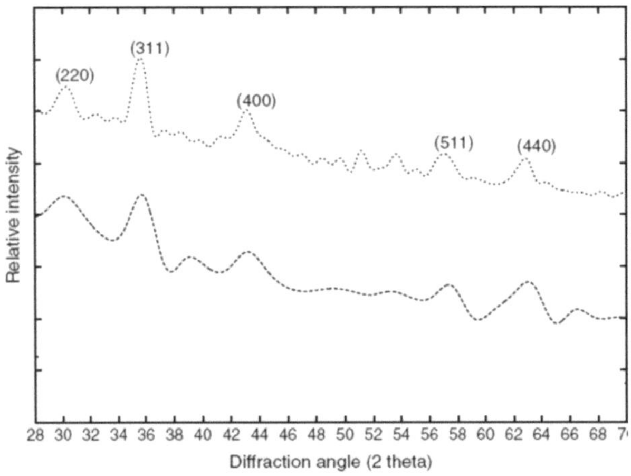

Fig. 5. XRD spectra of magnetic core-shell nanoparticles prepared by the oxidative polymerization of pyrrole monomer and magnetic ferrofluid

XRD analysis can add more information for understanding the nanostructure of the magnetite surrounded by the PPY shell. The studies made on this samples presented in the figure 5, is that the diminution of the number of atoms from the first and the second coordination shells of Fe atoms point out the existence of an electronic interaction between the magnetite nanoparticles and the surrounding PPY.
Another type of material which was investigated by X-ray diffraction is the magnetic core-shell nanoparticles based on $NiFe_2O_4$ and polypyrrole (figure 6). The XRD scan on this type of material shows a broad amorphous diffraction peak centred at around 2θ = 23^0, which correspond to the scattering from the polymer chain at the interplanar spacing of protonated

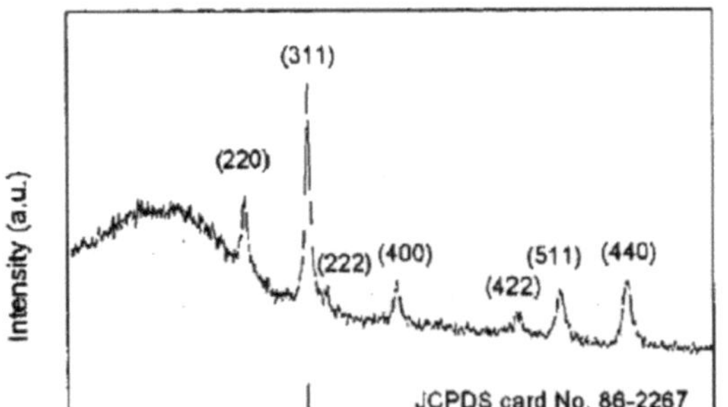

Fig. 6. XRD patterns of the magnetic core-shell nanoparticles based on polypyrrole and $NiFe_2O_4$

polypyrrole. The other seven diffraction peaks presented in the figure at $2\theta = 30.1^0, 35.6^0, 37.3^0, 43.4^0, 53.7^0, 57.4^0$ and 62.8^0 corresponding to the (220), (311), (222), (400), (422), (511) and (440) reflections of the ferrospinel $NiFe_2O_4$ are consistent with the reference standard data.

3.3 FTIR spectoscopy of the magnetic core-shell nanoparticles based on polypyrrole

Magnetic core-shell nanoparticles based on polypyrrole are materials which combine two components: an anorganic core and an organic shell, therefore the FTIR spectroscopy is one of the most important tool for structural investigation of this type of material.

Figure 7 compares the FTIR spectrum of a pure polypyrrole sample doped with DBS and the FTIR spectrum of magnetic core-shell nanoparticles based on functionalized polypyrrole with phenylalanine. The FTIR spectra of the nanocomposites contain the characteristic absorption bands of both constituents, namely, oxidized polypyrrole shell and Fe_3O_4. The intense absorption band located around 580 cm^{-1} is characteristic of Fe_3O_4. In the FTIR spectrum of magnetic core-shell nanoparticles the absorption bands which belongs to the polypyrrole chain at 914, 1190, 1465 cm^{-1} are shifted to higher wave numbers. The position of this bands are correlated with the conjugation length of the polypyrrole chain, the shift to higher frequencies indicates a decrease of the conjugation length of the functionalized polypyrrole chain in the functionalized magnetic nanoparticles as compared to unfunctionalized polypyrrole. The band presented at 1711 cm^{-1} ascribed the C=O bond, demonstrating that the function (phenylalanine) is attached on polypyrrole chain and incorporated in the magnetic core-shell nanoparticles (Nan et al., 2008).

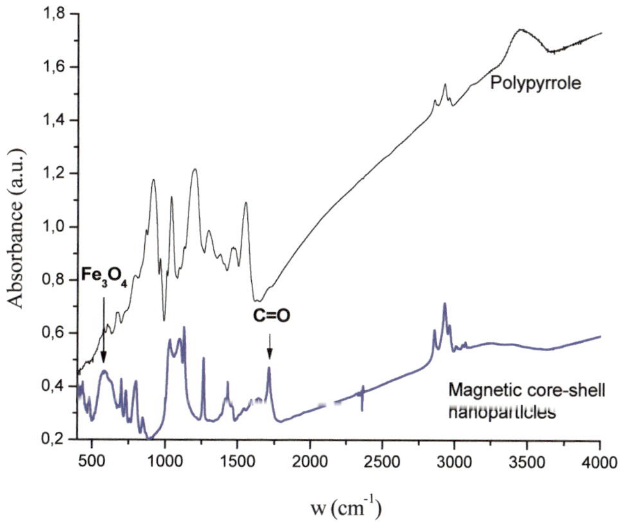

Fig. 7. The FTIR spectra of a pure polypyrrole sample doped with DBS and of the magnetic core-shell nanoparticles based on functionalized polypyrrole with phenylalanine

The FTIR spectra of polypyrrole and magnetic core-shell nanoparticles based on $ZnFe_2O_4$ and polypyrrole is presented in the figure 8. The fundamental vibration of pyrrole ring, peak observed at 1569 cm^{-1} in figure 2(a), has shifted to higher wavenumbers compared with 1543 cm^{-1} of pure polypyrrole. This may be due to the interaction between PPy and $ZnFe_2O_4$ nanoparticles and influence the skeletal vibrations, consequently delocalizing π-electrons. The peaks at 1289 cm^{-1} and 1046 cm^{-1} correspond to C-H in-plane vibration, while C-N stretching vibration and C–C out-of-plane ring deformation vibration are found at 1217 cm^{-1} and 931 cm^{-1}, respectively. C–H ring out-of-plane bending mode shows at 789 and 631 cm^{-1}. The appearance of peak at 1694 cm^{-1} is attributed to the overoxidization of polypyrrole. Additionally, a strong band at 572 cm^{-1} appears in the spectrum of $ZnFe_2O_4$ nanoparticles, which is assigned to the Fe–O stretching vibration mode. The conclusion which can be draw from the FTIR spectroscopoy in this case is that the $ZnFe_2O_4$ nanoparticles only serve as the nucleation sites for the polymerization of pyrrole because there is no chemical interaction between $ZnFe_2O_4$ and polypyrrole in the magnetic core-shell nanoparticles (Li et al., 2009).

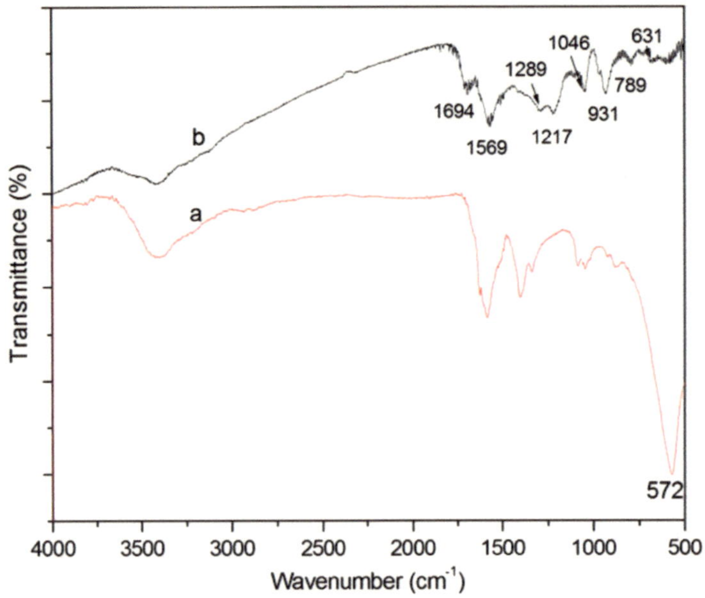

Fig. 8. FTIR spectra of (a) $ZnFe_2O_4$ nanoparticles and (b) magnetic core-shell nanoparticles based on $ZnFe_2O_4$ and polypyrrole nanoparticles

3.4 X-photoelectron spectroscopy of the magnetic core-shell nanoparticles based on polypyrrole

XPS is a surface-analysis method, consisting of the irradiation by X-rays of the investigated material, detection of the ejected photoelectrons, and their analysis by the kinetic energies (E_{kin}). The method provides information concerning the element composition (atomic concentration) of the surface, as well as the chemical state of the emitting atoms (valence states, oxidation degree, chemical ligands etc). The former information is inferred from the areas delimited by the photoelectron peaks. The latter one relates to the chemical shifts of the peaks with respect to the elemental state, induced by the chemical surrounding of the atoms. Chemical shift represents a change in E_b of a core electron of an element due to a change in chemical bonding of that element. Core binding energies are determined by electrostatic interaction between electron and the nucleus and are reduced by the electrostatic shielding of the nuclear charge from all other electrons in the atom. Removal or addition of electronic charge as a result of changes in bonding will alter the shielding: withdrawal of valence electron charge (oxidation) increase in E_b; addition of valence electron charge decrease in E_b.

Chemical shift information is a very powerful tool for functional group, chemical environment, oxidation state. According to the type of chemical bond and to the neighbor atoms, the binding energy of a given state can be shifted from a fraction of eV up to several eV. When several chemical bonds types are present, the spectrum peaks are splitted into several peaks which are sometimes almost merged. Using the appropriate software the peaks are deconvoluted into the components, each component corresponding to a particular bond type.

The high-resolution XPS spectra of C 1s, O 1s, N1s and Fe 2p core levels from the magnetic core-shell nanoparticles based on functionalized polypyrrole with cholesterol units, are given in the figure 9.

The spectrum of C 1s can be deconvoluted into three peaks corresponding to carbon atoms from different groups. The higher binding energy component at 288.7 eV correspond to carbon atoms from the group O-C=O present in the structure of the lauric acid double layer coating the magnetite nanoparticles from the magnetic fluid. The component peak located at 286 eV is ascribed to carbon atoms from the groups C-N, C-O from the structure of substituted pyrrole (see figure 9). The component at 285 eV corresponds to C-C, CH_n groups.

The oxygen spectrum in the figure 1 exhibit three components assigned to the oxygen atoms from C-O-C , O-C=O groups from the structure of the substituted pyrrole and lauric acid, respectively and from Fe-O located at 530 eV.

The N1s core-level spectrum, figure 1 shows four components ascribed to the nitrogen atoms in different chemical states characteristic for unsubstituted and substituted pyrrole structures (Kang et al., 1993): the component at 398 eV corresponds to C=N group, the more intense component located at 400 eV is ascribed to N-H, C-N groups, the component at 402 eV is attributed to the positively charged nitrogen N^+-H from unsubstituted pyrrole ring and the high binding energy component 403 eV ascribed to N-N group from the substituted pyrrole.

The Fe 2p spectrum contains the doublet Fe 2p3/2 and Fe 2p1/2 with binding energy values of 710 and 723.6 eV, typical for magnetite (Bhargava et al., 2007). Each peak from Fe 2p spectrum can be deconvoluted into two components corresponding to Fe^{3+} and Fe^{2+} ions from magnetite. One can observe the contribution of the Fe 2p3/2 satellites located at 714.19 eV and 719.6 eV which correspond to Fe^{2+} and Fe^{3+} species (Brundle et al., 1977).

3.5 Magnetic properties of the magnetic core-shell nanoparticles based on polypyrrole

The most important property of these magnetic core-shell nanoparticles based on polypyrrole is the magnetic properties. The magnetic properties for almost all type of magnetic nanoparticles are determined by vibrating sample magnetization (VSM) at theroom temperature. The magnetic properties of magnetic core-shell nanoparticles based on polypyrrole depend very much on the type of core but also on the polymerization condition applied for pyrrole monomer.

The magnetic curves for $ZnFe_2O_4$ nanoparticles and magnetic core-shell nanoparticles based on polypyrrole and $ZnFe_2O_4$ are presented in the figure 10, the authors determined that the magnetization saturation value Ms for magnetic core-shell nanoparticles based on $ZnFe_2O_4$ and polypyrrole is 17.8 emu/g, and the coercivity values Hc is about 130 Oe. However, the magnetization saturation value Ms of $ZnFe_2O_4$ nanoparticles is 31.92 emu/g, and the coercivity value Hc is 94 Oe. These results can be attributed to the fact that non-magnetic

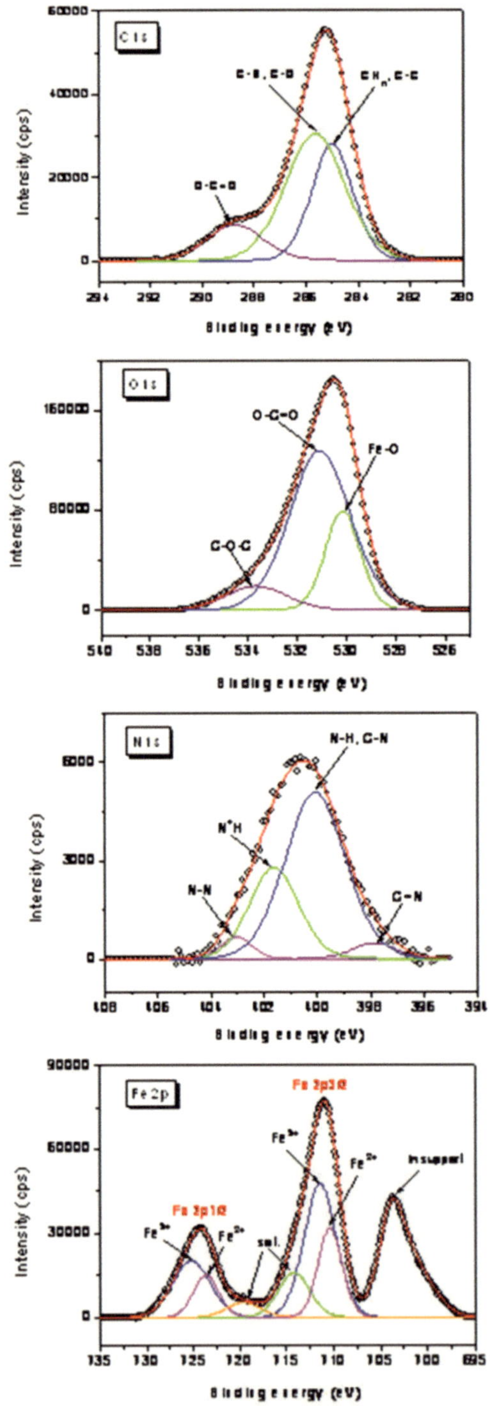

Fig. 9. High resolution XPS spectra C 1s, O 1s, N1s and Fe 2p core levels from the magnetite nanoparticles based on functionalized polypyrrole with cholesterol unites

polypyrrole coating layer on the surface of magnetic particles can decrease the magnetism of magnetic materials, while coercivity of nanocomposites increases due to the surface anisotropy upon coating and external morphology transformation (Li et al., 2009).

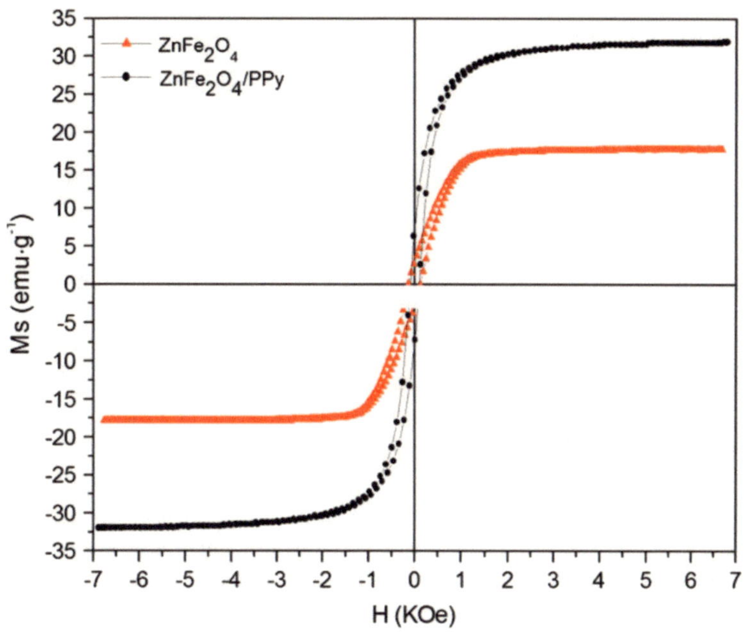

Fig. 10. Magnetic hysteresis loops of $ZnFe_2O_4$ nanoparticles and magnetic core-shell nanoparticles based on polypyrrole and $ZnFe_2O_4$ ($ZnFe_2O_4$/PPy).

In the case of magnetic nanoparticles based on functionalized polypyrrole prepared starting from magnetic ferrofluid the temperature magnetization show no hysteresis loop. For exemplification in the figure 11 we show the behavior of the magnetization at room temperature for magnetic core-shell functionalized polypyrrole with valine. The magnetization curves at room temperature of all investigated samples did not show hysteresis loops, thus proving the superparamagnetic behavior of the particles, the magnetization saturation value Ms is 63 emu/g. For such systems, the magnetic moment of the particle is free to rotate in response to the applied magnetic field when the blocking temperature is exceeded (Nan et al., 2008).

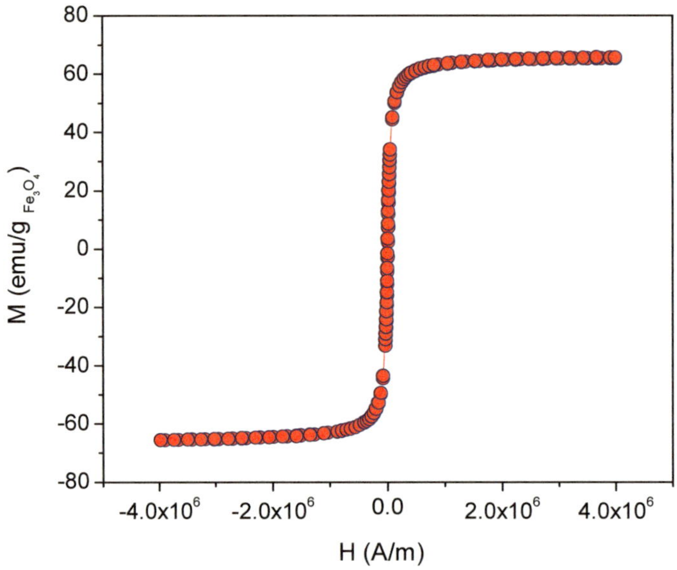

Fig. 11. Magnetization vs. applied magnetic field at room temperature of magnetic core-shell nanoparticles based on functionalized polypyrrole with valine

4. Conclusion

The magnetic core-shell nanoparticles based on polypyrrole in most cases are obtained via chemical oxidative polymerization in the presence of different types of magnetic nanoparticles. The future applications of this kind of materials requires a deep understanding of the nanostructure-properties relationship which implies a complex physical chemical characterization correlated with the synthesis parameters.

This review focused on polypyrrole which is one of the most used conducting polymers for applications due to its high electrical conductivity, relatively good stability and the easy polymerization by low cost and environmentally friendly methods.

The combination of conducting polypyrrole with different magnetic nanoparticles is a successful way to provide composites materials with improved processability and greater potential applicability. The nature of the component materials and the synthesis parameters allow the control of some magnetic core-shell nanoparticles properties such as colloidal stability, electrical conductivity, and magnetic susceptibly.

Due to the discovery of new tools for an easy functionalization of polypyrrole simultaneously the functionalization of magnetic nanoparticles a large utilization of this materials are expected in the future for many applications as: biomolecular recognition, diagnosis, organocatalysis and separation of biological material.

5. Acknowledgment

We greatly acknowledge to Dr. Doina Bica† and Dr. Ladislau Vekas from *Romanian Academy, Timisoara Branch, Magnetic Fluids Laboratory,* for ferrofluids samples preparation. This work was supported financially by CNCSIS -UEFISCSU, project number PNII - IDEI code 76/2010, Surface and interface science: physics, chemistry, biology, applications and project POS CCE ID 550/ SMIS-CSNR 12025, contract financing: nr. 235/16.08.2010.

6. References

Aldea, N.; Turcu, R.; Nan, A.; Craciunescu, I.; Pana, O.; Yaning, X.; Wu, Z.; Bica, D.; Vekas, L. & Matei, F. (2009). Investigation of nanostructured Fe_3O_4 polypyrrole core-shell composites by X-ray absorbtion spectroscopy and X-ray diffraction using synchrotron radiation. *Journal of Nanoparticle Research*, Vol.11, No.6, (August 2009), pp. 1429–1439, ISSN: 1388-0764

Aldissi, M.; Armes, S. P. (1991). Colloidal dispersions of conducting polymers. *Progress in Organic Coatings*, Vol.19, No.1, (April 1991), pp. 21-58, ISSN: 0300-9440

Armes, S. P.; Miller, J. F. & Vincent, B. (1987). Aqueous dispersions of electrically conducting monodisperse polypyrrole particles. *Journal of Colloid and Interface Science*, Vol.118, No.2, (August 1987), pp. 410-416, ISSN: 0021-9797

Armes, S. P. & Vincent, B. (1987). Dispersions of electrically conducting polypyrrole particles in aqueous media. *Journal of the Chemical Society, Chemical Communication*, (January 1987), No.4, pp. 288-290, ISSN: 0022-4936

Azioune, A.; Pech K.; Saoudi, B.; Chehimi, M. M.; McCarthy G. P. & Armes, S. P. (1999). Adsorption of human serum albumin onto polypyrrole powder and polypyrrole-silica nanocomposites. *Synthetic Metals*, Vol.102, No.1-3, (June 1999), pp. 1419-1420, ISSN: 0379-6779

Azioune, A.; Slimane, A. B.; Hamou, L. A.; Pleuvy, A.; Chehimi, M. M.; Perruchot, C. & Armes, S. P. (2004). Synthesis and characterization of active ester-functionalized polypyrrole -silica nanoparticles: application to the covalent attachment of proteins. *Langmuir*, Vol.20, No.8, (March 2004), pp. 3350-3356, ISSN: 0743-7463

Benabderrahmane, S.; Bousalem, S.; Mangeney, C.; Azioune, A.; Vaulay, M.-J. & Chehimi, M. M. (2005). Interfacial physicochemical properties of functionalized conducting polypyrrole particles. *Polymer*, Vol.46, No.4, (February 2005), pp. 1339-1346, ISSN: 0032-3861

Bhargava, G.; Gouzman, I.; Chun, C.M.; Ramanarayanan, T.A. & Bernasek, S.L. (2007). Characterization of the "native" surface thin film on pure polycrystalline iron: A high resolution XPS and TEM study. *Applied Surface Science*, Vol.253, No.9, (February 2007), pp. 4322-4329, ISSN: 0169-4332

Bhaumik, M.; Leswifi, T. Y.; Maity, A.; Srinivasu, V.V. & Onyango, M. S. (2011). Removal of fluoride from aqueous solution by polypyrrole/Fe_3O_4 magnetic nanocomposite. *Journal of Hazardous Materials*, Vol.186, No.1, (February 2011), pp. 150–159, ISSN: 0304-3894

Bica, D.; Vékás, L.; Avdeev, M. V.; Marinica, O.; Socoliuc, V.; Balasoiu, M. & Garamus, V. M. (2007). Sterically stabilized water based magnetic fluids: Synthesis, structure and properties. *Journal of Magnetism and Magnetic Materials*, Vol.311, No.1, (April 2007), pp. 17–21, ISSN: 0304-8853

Birsöz, B.; Baykal, A.; Sözeri, H. & Toprak, M. S. (2010). Synthesis and characterization of polypyrrole–$BaFe_{12}O_{19}$ nanocomposite. *Journal Alloys and Compounds*, Vol.493, No.1-2, (March 2010), pp. 481–485, ISSN: 0925-8388

Bjorklund, R. B. & Liedberg, B. (1986). Electrically conducting composites of colloidal polypyrrole and methylcellulose. *Journal of the Chemical Society, Chemical Communication*, No.16, (January 1986), pp. 1293-1295, ISSN: 0022-4936

Bousalem, S.; Yassar, A.; Basinska, T.; Miksa, B.; Slomkowski, S.; A. Azioune & M. M. Chehimi, (2003). Synthesis, characterization and biomedical applications of functionalized polypyrrole-coated polystyrene latex particles. *Polymers for Advanced Technologies*, Vol.14, No.11-12, (November-December 2003), pp. 820-825, ISSN: 1099-1581

Bousalem, S.; Mangeney, C.; Chehimi, M. M.; Basinska, T.; Miksa, B. & Slomkowski, S. (2004a). Synthesis, characterization and potential biomedical applications of N–succinimidyl ester functionalized, polypyrrole-coated polystyrene latex particles. *Colloid and Polymer Science*, Vol.282, No.12, (October 2004), pp. 1301-1307, ISSN: 1435-1536

Bousalem, S.; Mangeney, C.; Alcote, Y.; Chehimi, M. M.; Basinska, T. & Slomkowski, S. (2004b). Immobilization of proteins onto novel, reactive polypyrrole-coated polystyrene latex particles, *Colloids and Surfaces A: Physicochemical Engineering Aspects*, Vol.249, No.1-3, (November 2004), pp. 91–94, ISSN: 0927-7757

Bousalem, S.; Benabderrahmane, S.; Sang, Y. Y. C.; Mangeney, C. & Chehimi, M. M. (2005). Covalent immobilization of human serum albumin onto reactive polypyrrole-coated polystyrene latex particles. *Journal of Materials Chemistry*, Vol.15, No.30, (June 2005), pp. 3109-3116, ISSN: 0959-9428

Brundle, C. R.; Chuang, T.J. & Wandelt, K. (1977). Core and valence level photoemission studies of iron oxide surfaces and the oxidation of iron. *Surface Science*, Vol.68, (November 1977), pp. 459-468, ISSN: 00396028

Cairns, D. B. & Armes, S. P. (1999). Synthesis and characterization of submicrometer-sized polypyrrole-polystyrene composite Particles. *Langmuir*, Vol.15, No.23, (November 1999), pp. 8052-8058, ISSN: 0743-7463

Cawdery, N.; Obey, T. M. & Vincent, B. (1988). Colloidal dispersions of electrically conducting polypyrrole particles in various media. *Journal of the Chemical Society, Chemical Communication*, No.17, (January 1988), pp. 1189-1190, ISSN: 0022-4936

Deng, J.; Peng, Y.; He, C.; Long, X.; Li, P. & Chan, A. S. C. (2003). Magnetic and conducting Fe_3O_4–polypyrrole nanoparticles with core-shell structure. *Polymer International*, Vol.52, No.7, (July 2003), pp. 1182–1187, ISSN: 1097-0126

Feng, X.; Huang, H.; Ye, Q.; Zhu, J. J. & Hou, W. (2007), Ag/Polypyrrole Core-Shell Nanostructures: Interface Polymerization, Characterization, and Modification by Gold Nanoparticles. *The Journal of Physical Chemistry C*, Vol.111, No.24, (June 2007), pp. 8463-8468, ISSN: 1932-7447

Goller, M. I.; Barthet, C.; McCarthy, G.P.; Corradi, R.; Newby, B. P.; Wilson, S. A.; Armes, S. P. & Luk S. Y. (1998). Synthesis and characterization of surface-aminated polypyrrole-silica nanocomposites. *Colloid and Polymer Science*, Vol.276, No.11, (November 1998), pp. 1010-1018, ISSN: 1435-1536

Hao, L.; Gong, X.; Xuan, S.; Zhang, H.; Gong, X.; Jiang, W. & Chen, Z. (2006). Controllable fabrication and characterization of biocompatible core-shell particles and hollow capsules as drug carrier. *Applied Surface Science*, Vol.252, No.24, (October 2006), pp. 8724–8733, ISSN: 0169-4332

Huijs, F. M. & Lang, J. (2000). Morphology and film formation of poly(butyl methacrylate)-pyrrole core-shell latex particles. *Colloid and Polymer Science*, Vol.278, No.8, (August 2000), pp. 746-756, ISSN: 1435-1536

Huijs, F. M.; Lang, J.; Kalicharan, D.; Vercauteren, F. F.; Van Der Want, J. J. L. & Hadziioannou, G. (2001). Formation of transparent conducting films based on core-shell latices: influence of the polypyrrole shell thickness. *Journal of Applied Polymer Science*, Vol.79, No.5, (January 2001), pp. 900–909, ISSN: 1097-4628

Jiang, J.; Ai, L. H. & Zheng J.L. (2010). Preparation and characterization of polypyrrole/ferrospinel nanocomposites by W/O microemulsion pathway. *Journal of Macromolecular Science, Part B: Physics*, Vol.49, No.4, (June 2010), pp. 652-657, ISSN: 0022-2348

Karsten, S.; Ameen, M. A.; Kalläne, S. I.; Nan, A.; Turcu, R. & Liebscher, J. (2010). A versatile method of tethering biomolecules to pyrrole precursors for functionalized magnetic polypyrrole core-shell nanoparticles. *Synthesis*, No.17, pp. 3021-3028, ISSN: 0039-7881

Kang, E.T.; Neoh, K.G. & Tan, K.L. (1993). X-ray photoelectron spectroscopic studies of electroactive polymers. *Advances in Polymer Science*, Vol.106, pp.135-190, ISSN: 0065-3195

Lascelles, S. F. & Armes, S. P. (1995). Synthesis and characterization of micrometersized polypyrrole-coated polystyrene latexes. *Advanced Material*, Vol.7, No.10, (October 1995), pp. 864-866, ISSN: 1521-4095

Lascelles, S. F.; Armes, S. P.; Zhdan, P. A.; Greaves, S. J.; Brown, A. M.; Watts, J. F.; Leadley, S. R. & Luk, S. Y.(1997). Surface characterization of micrometre-sized, polypyrrole-coated polystyrene latexes: verification of a "core–shell" morphology. *Journal of Materials Chemistry*, Vol.7, No.8, pp. 1349-1355, ISSN: 0959-9428

Lascelles, S. F. & Armes, S. P. (1997). Synthesis and characterization of micrometre-sized,polypyrrole-coated polystyrene latexes. *Journal of Materials Chemistry*, Vol.7, No.8, pp. 1339-1347, ISSN: 0959-9428

Lee, J. M.; Lee, D. G.; Lee, S. J.; & Kim, J. H. (2009). One-step synthetic route for conducting core-shell poly(styrene/pyrrole) nanoparticles. *Macromolecules*, Vol.42, No.13, (July 2009), pp. 4511-4519, ISSN: 0024-9297

Li, J. L. & Liu, X. Y. (2008). Fabrication and biofunctionalization of selenium-polypyrrole core-shell nanoparticles for targeting and imaging of cancer cells. *Journal of Nanoscience and Nanotechnology*, Vol.8, No.5, (may 2008), pp. 2488–2491, ISSN: 1533-4880

Li, X.; Wan, M.; Wei Y.; Shen, J. & Chen, Z. (2006). Electromagnetic functionalized and core-shell micro/nanostructured polypyrrole composites. *The Journal of Physical Chemistry B*, Vol.110, No. 30, (August 2006), pp. 14623-14626, ISSN: 1520-6106

Li, Y.; Yi, R.; Yan, A.; Deng, L.; Zhou, K. & Liu, X. (2009). Facile synthesis and properties of $ZnFe_2O_4$ and $ZnFe_2O_4$/polypyrrole core-shell nanoparticles. *Solid State Sciences*, Vol.11, No.8, (August 2009), pp. 1319–1324, ISSN: 1293-2558

Liu, X.; Wu, H.; Renb, F.; Qiu G. & Tang, M. (2008). Controllable fabrication of SiO_2/polypyrrole core–shell particles and polypyrrole hollow spheres. *Materials Chemistry and Physics*, Vol.109, No.1, (May 2008), pp. 5–9, ISSN: 0254-0584

Lu, Y.; Pich, A. & Alder H. (2003). Synthesis and Characterization of nanometer-sized polypyrrole composites. *Synthetic Metals*, Vol.135-136, (April 2003), pp. 37-38, ISSN: 0379-6779

Maeda, S. & Armes, S. P. (1994). Preparation and characterisation of novel polypyrrole-silica colloidal nanocomposites. *Journal of Materials Chemistry*, Vol.4, No.6, pp.935-942, ISSN: 0959-9428

Maeda, S. & Armes, S. P. (1995). Preparation and Characterization of Polypyrrole-Tin(IV) Oxide Nanocomposite Colloids. *Chemistry Materials*, Vol.7, No.1, (January 1995), pp. 171-178, ISSN: 0897-4756

Mangeney, C.; Bousalem, S.; Connan, C.; Vaulay, M.-J.; Bernard, S.; & Chehimi, M. M. (2006). Latex and Hollow Particles of Reactive Polypyrrole: Preparation, Properties, and Decoration by Gold Nanospheres. *Langmuir*, Vol.22, No.24, (November 2006), pp.10163-10169, ISSN: 0743-7463

Mangeney, C.; Fertani, M.; Bousalem, S.; Zhicai, M.; Ammar, S.; Herbst, F.; Beaunier, P.; Elaissari, A. & Chehimi, M. M. (2007). Magnetic Fe_2O_3-polystyrene/ppy core/shell particles: bioreactivity and self-assembly. *Langmuir*, Vol.23, No.22, (October 2007), pp. 10940-10949, ISSN: 0743-7463

Marini, M.; Pilati, F. & Pourabbas, B. (2008). Smooth surface polypyrrole-silica core-shell nanoparticles: preparation, characterization and properties. *Macromolecular Chemistry and Physics*, Vol.209, No.13, (July 2008), pp.1374-1380, ISSN: 1022-1352

Masuda, H.; Tanaka, S. & Kaeriyama, K. (1989). Soluble Conducting Polypyrrole: Poly(3-octylpyrrole). *Journal of the Chemical Society, Chemical Communication*, No.11, (January1989), pp. 725-726, ISSN: 0022-4936

McCarthy, G. P. & Armes, S. P. (1997). Synthesis and characterization of carboxylic acid-functionalized polypyrrole-silica microparticles using a 3-substituted pyrrole comonomer. *Langmuir*, Vol.13, No.14, (July 1997), pp. 3686-3692, ISSN: 0743-7463

Nan, A.; Karsten, S.; Craciunescu, I.; Turcu, R.; Vekas, L. & Liebscher, J. (2008). New shells for magnetic nanoparticles based on polypyrrole functionalized with α-amino acids. *ARKIVOC*, Vol.xv, pp. 307-320, ISSN: 1551-7004

Nan, A.; Turcu, R.; Bratu, I.; Leostean, C.; Chauvet, O.; Gautron, E. & Liebscher J. (2010). Novel magnetic core-shell Fe_3O_4 polypyrrole nanoparticles functionalized by peptides or albumin. *ARKIVOC*, Vol.x, pp. 185-198, ISSN: 1551-7004

Nguyen, M. T. & Diaz, A. F. (1994). A novel method for the preparation of magnetic nanoparticles in a polypyrrole powder. *Advanced Materials*, Vol.6, No.11, pp. 858-860, (November 1994), ISSN: 1521-4095

Pana, O.; Teodorescu, C.M.; Chauvet, O.; Payen, C.; Macovei, D.; Turcu, R.; Soran, M.L.; Aldea, N. & Barbu, L. (2007). Structure, morphology and magnetic properties of Fe–Au core-shell nanoparticles. *Surface Science*, Vol.601, No.18, (September 2007), pp. 4352–4357, ISSN: 00396028

Park, D. H.; Oh, J. M.; Shul, Y. G. & Choy, J. H. (2008). Fe_3O_4@polypyrrole core–shell nanohybrid for efficient DNA retrieval. *Journal of Nanoscience and Nanotechnology*, Vol.8, No.10, (October 2008), pp. 5014–5017, ISSN: 1533-4880

Pourabbasa, B. & Pilati, F. (2010). Polypyrrole grafting onto the surface of pyrrole-modified silica nanoparticles prepared by one-step synthesis. *Synthetic Metals*, Vol.160, No.13-14, (July 2010), pp. 1442-1448, ISSN: 0379-6779

Qiu, G.; Wang, Q. & Nie, M. (2006). Polypyrrole-Fe_3O_4 magnetic nanocomposite prepared by ultrasonic irradiation. *Macromolecular Materials and Engineering*, Vol.291, No.1, (January 2006), pp. 68–74, ISSN: 1439-2054

Rojas, D. M.; Sole´, J. O.; Ayyad, O. & Romero, P. G. (2008). Facile one-pot synthesis of self-assembled silver@polypyrrole core/shell nanosnakes. *Small*, Vol.4, No.9, (September 2008), pp. 1301–1306, ISSN: 1613-6829

Stanke, D.; Hallensleben, M. L. & Toppare L. (1993). Electrically conductive poly(methyl methacrylate-g-pyrrole) via chemical oxidative polymerization. *Synthetic Metals*, Vol.55, No.2-3, (March 1993), pp. 1108-1113, ISSN: 0379-6779

Turcu, R.; Bica, D.; Vekas, L.; Aldea, N.; Macovei, D.; Nan, A.; Pana, O.; Marinica, O.; Grecu, R.& Pop, C.V.L. (2006). Synthesis and characterization of nanostructured polypyrrole-magnetic particles hybrid material. *Romanian Reports in Physics*, Vol.58, No.3, pp. 359–367, ISSN: 1221-1451

Turcu, R.; Pana, O.; Nan, A.; Craciunescu, I.; Chauvet, O. & Payen, C. (2008). Polypyrrole coated magnetite nanoparticles from water based nanofluids. *Journal of Physics D: Applied Physics*, Vol.41, No.24, (December 2008), 245002 (9pp), ISSN: 0022-3727

Vasilyeva, S. V.; Vorotyntsev, M. A.; Bezverkhyy, I.; Lesniewska, E.; Heintz, O. & Chassagnon, R. (2008). Synthesis and characterization of palladium nanoparticle/polypyrrole composites. *The Journal of Physical Chemistry C*, Vol.112, No.50, (December 2008), pp. 19878–19885, ISSN: 1932-7447

Wang, J.; Sun, L.; Mpoukouvalas, K.; Lienkamp, K.; Lieberwirth, I.; Fassbender, B.; Bonaccurso, E.; Brunklaus, G.; Muehlebach, A.; Beierlein, T.; Tilch, R.; Butt, H. J. & Wegner, G. (2009). Construction of redispersible polypyrrole core–shell nanoparticles for application in polymer electronics. *Advanced Materials*, Vol.21, No.10-11, (March 2009), pp.1137-1141, ISSN: 1521-4095

Wang, S. & Shi, G. (2007). Uniform silver/polypyrrole core-shell nanoparticles synthesized by hydrothermal reaction. *Materials Chemistry and Physics*, Vol.102, No.2-3, (April 2007), pp. 255–259, ISSN: 0254-0584

Wu, T. M.; Yen, S. J.; Chen, E. C.; Sung, T. W. & Chiang, R. K. (2007). Conducting and magnetic behaviors of monodispersed iron oxide/polypyrrole nanocomposites synthesized by in situ chemical oxidative polymerization. *Journal of Polymer Science: Part A: Polymer Chemistry*, Vol. 45, No.20, (October 2007), pp. 4647–4655, ISSN: 1099-0518

Wuang, S. C.; Neoh, K. G.; Kang, E. T.; Pack, D. W. & Leckband, D. E. (2007). Synthesis and functionalization of polypyrrole-Fe_3O_4 nanoparticles for applications in biomedicine. *Journal of Materials Chemistry*, Vol.17, No.31, pp. 3354-3362, ISSN: 0959-9428

Xing, S.; Tan, L. H.; Yang, M.; Pan, M.; Lv, Y.; Tang, Q.; Yang, Y. & Chen H. (2009). Highly controlled core/shell structures: tunable conductive polymer shells on gold nanoparticles and nanochains. *Journal of Materials Chemistry*, Vol.19, No.20, pp. 3286–3291, ISSN: 0959-9428

Xu, P.; Han, X.; Wang, C.; Zhao, H.; Wang, J.; Wang, X. & Zhang, B. (2008). Synthesis of electromagnetic functionalized barium ferrite nanoparticles embedded in polypyrrole. *The Journal of Physical Chemistry B*, Vol.112, No.10, (March 2008), pp. 2775-2781, ISSN: 1520-6106

Yang, X.; Dai, T. & Lu, Y. (2006). Synthesis of novel sunflower-like silica/polypyrrole nanocomposites via self-assembly polymerization. *Polymer*, Vol.47, No.1, (January 2006), pp. 441–447, ISSN: 0032-3861

Ye, S. & Lu, Y. (2008). Optical Properties of Ag@Polypyrrole Nanoparticles Calculated by Mie Theory. *The Journal of Physical Chemistry C*, Vol.112, No.24, (June 2008), pp. 8767-8772, ISSN: 1932-7447

Ye, S.; Fang L. & Lu, Y. (2009). Contribution of charge-transfer effect to surface-enhanced IR for Ag@PPy nanoparticles. *Physical Chemistry Chemical Physics*, Vol.11, No.14, pp. 2480–2484, ISSN: 1463-9076

Yip, Y.; Benabderrahmane, S.; Zhicai, M.; Bousalem, S.; Mangeney C. & Chehimi, M. M. (2006). Interactions of reactive polypyrrole-coated polystyrene latex particles with gold nanoparticles and silanized glass. *Surface and Interface Analysis*, Vol.38, No.4, (April 2006), pp. 535-538, ISSN: 1096-9918

Zhang, C.; Li, Q. & Ye, Y. (2009). Preparation and characterization of polypyrrole/nano-$SrFe_{12}O_{19}$ composites by in situ polymerization method. *Synthetic Metals*, Vol.159, No.11, (June 2009), pp. 1008–1013, ISSN: 0379-6779

Zhang, H.; Zhong, X.; Xu, J. J. & Chen, H. Y. (2008). Fe_3O_4/polypyrrole/Au nanocomposites with core/shell/shell structure: synthesis, characterization, and their electrochemical properties. *Langmuir*, Vol.24, No.23, (December 2008), pp. 13748-13752, ISSN: 0743-7463

Part 5

Membranes

9

Ion Conductive Polymer Electrolyte Membranes and Fractal Growth

Shahizat Amir, Nor Sabirin Mohamed and Siti Aishah Hashim Ali
University of Malaya
Malaysia

1. Introduction

It has been widely accepted that Euclidean geometry plays an important role in shaping the way natural forms are viewed in science and mathematics, arts and even the human psyche (Hastings & Sugihara, 1993). This happens because man always seeks to find simplicity and order in nature, and often makes approximation on natural forms that may be essentially complex and irregular. Hence, leaves are roughly ellipses, planets are spheres and spruce trees are cone-shaped. However, shapes such as coastlines, fern leaves and clouds are not easily described by traditional Euclidean geometry. Nevertheless, they often possess a remarkable invariance under changes of magnification. With a certain scale of magnification, the pattern is seen as repeating itself. Since the term 'fractal' was first coined by Mandelbrot (Mandelbrot, 1983), study of fractals has increasingly become an interest for scientists and mathematicians. Consequently many researchers study the growth and shapes of fractals through theoretical modeling and computer simulations of fractal patterns.
Simulation model of fractal patterns found in polymer electrolyte membranes provides another interesting perspective in the study of ion conductive polymer membranes. The characteristics and scientific aspects of the model have been studied and computer program s to simulate the growth of the patterns have been developed. Fractal aggregates especially diffusion-limited aggregate involve the random walk of particles and their subsequent sticking (Chandra & Chandra, 1994). To obtain fractal aggregates in laboratory framework, a system with particles in random walk is required. In most polymer electrolytes, the anions as well as the cations are found to be mobile and thus can be considered as a natural framework for fractal growth. The polymers act as a host while the inorganic salts dissociate in them to provide the ions necessary for conduction. According to Chandra (1996), fractals formed in the PEO-NH_4I polymer electrolyte films are principally due to the random walk and subsequent aggregation of iodine ions. In other research as well, Fujii et al. (1991) have successfully carried out fractal dimension calculations of dendrite, of fractal patterns observed on the surface of a conducting polymer polypyrrole, after an 'undoping' process. Recent studies of fractals in polymers that involved modeling and/or simulation include Janke & Schakel (2005), Lo Verso et al. (2006) and Marcone et al. (2007). On the other hand, Rathgeber et al. (2006) have done some work on theoretical modeling and experimental studies of dendrimers. There have also been experimental studies of crystal pattern transition from dendrites through fourfold-symmetric structures to faceted crystals of ultra thin poly(ethylene oxide) films which were carried out by Zhang et al. (2008). These research

works on fractals were done only on laboratory experiments, theoretical modeling and experimental studies, or modeling and computer simulations. Most recently, Amir et al. (2010a; 2011a) succeeded in integrating all the three approaches; experimental, modeling and simulation in studying fractals in different ion conductive polymer membranes (Amir et al., 2010a; 2010c; 2011a; 2011b). The study of fractal growth on ion conductive polymer membranes is useful in understanding the movement of ions in the films and can also be used to study heavy metal accumulation in diseased glands in humans and fishes (Chandra, 1996).

2. An overview of fractals

Benoit B. Mandelbrot (Mandelbrot, 1983) introduced the term 'fractal' that refers to a family of complex geometrical that can be characterized by a fractional or non-integer dimensionality. The concept of fractals has attracted the interest of scientists in many fields (Feder, 1988). A huge number of papers related to the word 'fractal' has been published, spanning fields ranging from physical geometry, such as surface structure of sea beds (Golubev et al., 1987), non-equilibrium growth phenomena (Shibkov et al., 2001) and distribution of intervals between earthquakes (Dargahi-Noubary, 1997), to ecology that involves fungal structure (Tordoff et al., 2007) and power law relationship between the area of a quadrate and the structure of peat systems (Sławinski et al.,2002). Even in cosmology with the study of the structure of star clusters and galaxies, the big bang theory of the origin of the universe and also in developmental biology portrayed by lung branching patterns, heart rhythms and structure of neurons (Hastings & Sugihara, 1993).

The most amazing thing about fractal is the variety of its applications. Besides theoretical applications, it can be used to compress data in the Encarta Encyclopedia and to create realistic landscapes in movies like Star Trek. The places where fractals can be found include almost every part of the universe, from bacteria cultures to galaxies and to human body. Many studies of fractals related to fields such as astronomy (Combes, 1998), biology (Stanley et al., 1994) and chemistry (Villani & Comenges, 2000). In mathematics, the study of fractals revolves around data compression, fractal art and diffusion.

Many of fractal growth models were also found to be suitable with experimental studies of electrochemical electrodeposition (Barkey, 1991), electrochemical polymerization (Kaufmann et al., 1987) and DLA growth structures of many metal aggregates in the presence of a magnetic field as external stimuli (Okubo et al., 1993). The formation of fractals without using any external stimuli has been reported by Chandra & Chandra (1993); Mohamed & Arof (2001) and Amir et al. (2010a; 2010c; 2011a; 2011b).

2.1 Fractal geometry

Fractal or fractional dimension is something that can never be understood inside the realm of elementary geometry. It is another field in which at least one of Euclid's postulates does not hold, and where other mathematical realities emerge. Thus, it can be said that there are two types of geometry: Euclidean and non-Euclidean geometries. In the first group, are the plane geometry, solid geometry, trigonometry, descriptive geometry, projective geometry, analytical geometry and differential geometry. In the second group, there are hyperbolic geometry, elliptic geometry and fractal geometry. Almost all geometric forms used for building man made objects belong to Euclidean geometry. They compromised of lines, planes, rectangular volumes, arcs, cylinders, spheres and defined shapes. These elements can be classified as belonging to an integer dimension: 1, 2, or 3. Table 1 gives the summary of the major differences between fractal and the traditional Euclidean geometry.

EUCLIDEAN	NON EUCLIDEAN (FRACTAL)
Traditional (>2000 yrs)	Modern monsters (~ 30 yrs)
Based on characteristic size or scale	No specific scaling
Suits man made objects	Appropriate for natural shapes
Describes by formula	(Recursive) algorithm

Table 1. A comparison of Euclidean and fractal geometry (Peitgen & Saupe, 1988)

Fractal geometry allows length measurements to change in a non-integer or fractional way when the unit of measurements changes. The governing exponent D is called *fractal dimension* (Smith et al., 1990). The fractal dimension is a statistical quantity that gives an indication of how completely a fractal appears to fill space, as one zooms down to finer and finer scales. Fractal object has a property that more fine structure is revealed as the object is magnified, similarly like morphological complexity, which means that more fine structure (increased resolution and detail) is revealed with increasing magnification. Fractal dimension measures the rate of addition of structural detail with increasing magnification, scale or resolution. The fractal dimension, therefore, serves as a *quantifier of complexity*.

2.1.1 Self similarity

The main idea behind fractal geometry is self similarity. Self-similarity means that a structure (or process) can be decomposed into smaller copies of itself. This means that a self-similar structure is infinite. Self-similarity entails scaling. For an observable $A(x)$, which is a function of a variables x: $A = A(x)$, obeys a scaling relationship:

$$A\,(\lambda x) = \lambda^{s} A(x) \tag{1}$$

where λ is a constant factor and s is the scaling exponent, which is independent of x. For example, in a three-dimensional Euclidean space, volume scales as the third power of linear length, whereas fractals according to their fractal dimension (Focardi, 2003). Approximate self-similarity means that the object doesn't display perfect copies of itself. For example a coastline is a self-similar object, a natural fractal, but it does not have perfect self-similarity. A map of a coastline consists of bays and headlands, but when magnified, the coastline isn't identical but statistically the average proportions of bays and headlands remain the same no matter the scale (Judd, 2003).

It is not only natural fractals that display approximate self-similarity, the Mandelbrot set is another example. Identical pictures do not appear straight away, but when magnified, smaller examples will appear at all levels of magnification (Judd, 2003). *Statistical self-similarity* means that the degree of complexity repeats at different scales instead of geometric patterns. Many natural objects are statistically self-similar where as artificial fractals geometrically self-similar (Yadegari, 2003).

Geometrical similarity is a property of the space-time metric, whereas physical similarity is a property of the matter fields. The classical shapes of geometry do not have this property; a circle if on a large enough scale will look like a straight line. This is why people believed that the world was flat, the earth just looks that way to humans (Carr & Coley, 2003).

2.1.2 Fractal dimension

Fractal dimension is a measure of how complicated a self-similar figure is. In a rough sense, it measures how many points lie in a given set. The fractal dimension is often fractional. However, in algebra, the dimension of a space is defined as the smallest number of vectors needed to span that space (Rucker, 1984). In the 3 dimensional space, mathematicians traditionally denote the coordinates of three orthonormal vectors x, y and z. But sets are usually not vector spaces. Nevertheless, for aggregates, a fractal dimensionality in terms of scaling relationship between two different aggregate's properties X and Y (e.g. mass and length) can be observed such as (Meakin, 1988):

$$Y \propto X\ d_f \quad (2)$$

where d_f is all purpose fractal dimension as described by Meakin (1988).
Mandelbrot (1983) developed the 'concept of homothetic dimension' relative to geometric fractals. Let X be a complete metric space and let $A \subset X$. If N (A, ϵ) is the least number of balls of radius less than ϵ that are needed to cover A, then the number D (A) defined by

$$D(A) = \lim_{\varepsilon \to 0} \frac{\ln N(A,\varepsilon)}{\ln \frac{1}{\varepsilon}} \quad (3)$$

and is called the fractal dimension of A.
For each part (N) of the fractal deducted from the whole and having a homothetic ratio $r(N)$, the fractal dimension d_f is defined as:

$$d_f = \frac{Log(N)}{Log\left(\frac{1}{r}\right)} \quad (4)$$

For example, the Von Koch's snowflake iteration as illustrated in Figure 1, each side of unit 1 of a triangle is divided by 3, hence. $r = 1/3$. The central third of one side is replaced by 2 smaller lines of length 1/3. Therefore, one line is now subdivided in 4 smaller lines of length 1/3, hence. $N = 4$. Its fractal dimension now becomes:

$$d_f = \frac{Log(4)}{Log(3)} \approx 1.262 \quad (5)$$

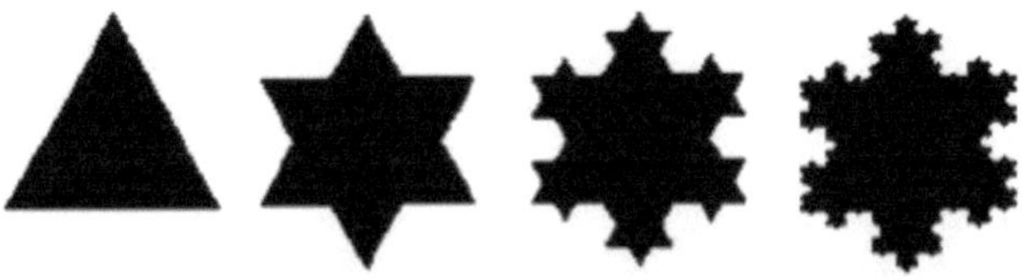

Fig. 1. Construction of the Von Koch's snowflake

2.1.3 Types of fractals

Fractal geometry is the geometry of structures that have a scaling symmetry. The simplest types of fractals are self-similar fractals that are invariant to an isotropic change of length scale (Meakin, 1991). Another approach to fractals is the way they are generated, for example by an *iterative* process. This process of iteration leads to different categories of fractals. Generally fractals can be divided into two main categories:

1. Deterministic Fractals
2. Random Fractals

2.1.3.1 Deterministic fractals

Deterministic fractals are generated by an iterative process. The term deterministic means that a simple process of iteration is applied to build the fractals such as the iteration of a complex function that generates the 'Mandelbrot Set' as shown in Figure 2. The iteration process is a geometrical transformation called *generator* on an object. This object is called *initiator*. For the construction of the so-called 'Koch's Curve' the transformation for each iteration is repeated. To build this fractal, a line of unit 1 is divided by 3 and the central $\frac{1}{3}$ is taken out and is replaced by 2 lines of length $\frac{1}{3}$. On the next iteration, the same transformation is applied on the remaining lines repeatedly. Its construction is described in Figure 2 as follows:

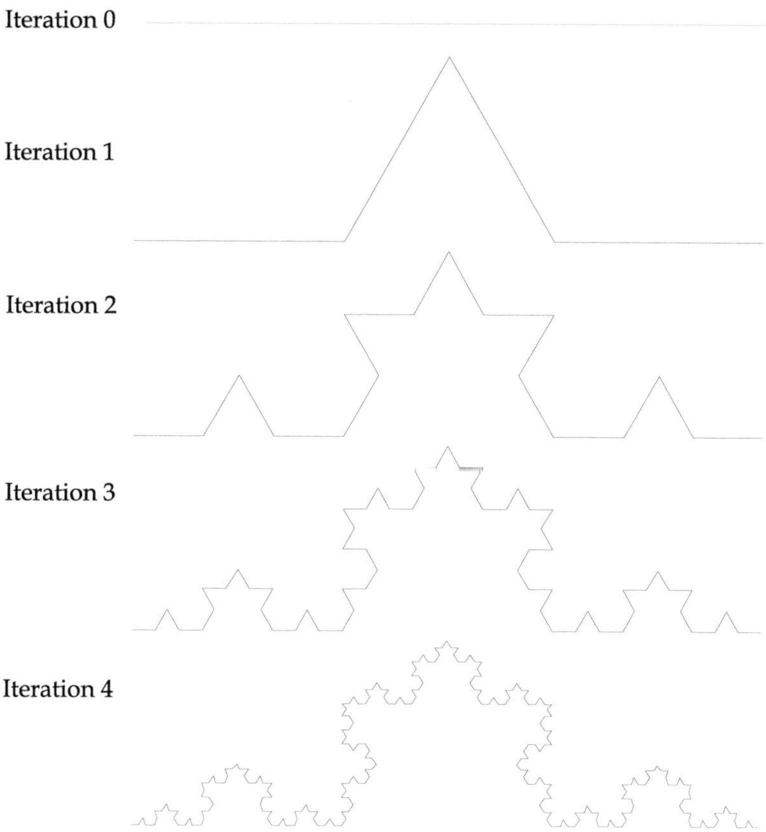

Fig. 2. Construction of Koch's curve (Addison, Paul S., 1997)

An important property of this fractal is its length that is infinity. The length of the initiator is 1, therefore, after the first iteration; the calculated length of the object is 4 lines of length $\frac{1}{3}$, that is $\frac{4}{3}$. Then the second iteration gives 16 lines of length $\frac{1}{9}$. The length now becomes equal to $\frac{16}{9}$. More generally, at each iteration n, the length becomes equal to $(4/3)^n$. As n tends to infinity, the length is $(4/3)^\infty = \infty$. The property of self-similarity can also be easily seen, as illustrated in Figure 3.

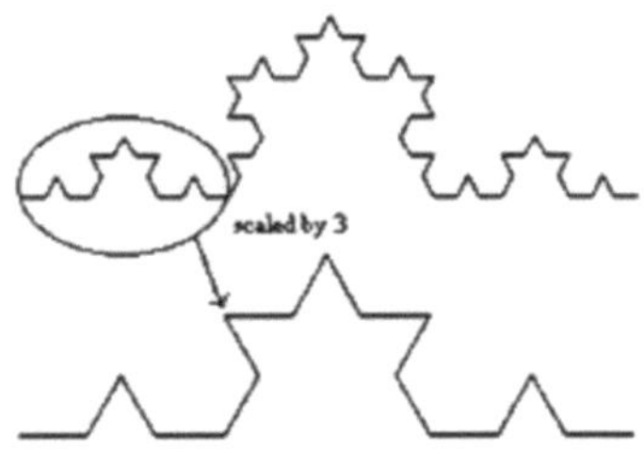

Fig. 3. Self similarity property of the Koch Curve

2.1.3.2 Random fractals

Random fractals are generated by stochastic processes, for example, trajectories of the Brownian motion, Lévy flight, fractal landscapes and the Brownian tree. The latter yields the so-called mass- or dendritic fractals, for example, diffusion-limited aggregation clusters. In the 1980's, Meakin developed different aggregation models in order to study the various ways an aggregate could be generated (Meakin, 1988; Meakin, 1991). Those aggregation models which are similar to the L-system are computer-generated where a set of transformation is applied on the generator that, in this case, would be an initial particle or cluster in the model. Random fractals have been used extensively in computer graphics to model natural objects (Ebert, 1996).

Many attractive images and life-like structures can be generated using models of physical processes from areas of chemistry and physics. One such example is diffusion limited aggregation (DLA) which describes, among other things, the diffusion and aggregation of zinc ions in an electrolytic solution onto electrodes. 'Diffusion' is because the particles forming the structure wander around randomly before attaching themselves (aggregating) to the structure. 'Diffusion-limited' because the particles are considered to be in low concentrations so they do not come in contact with each other and the structure grows one particle at a time rather then by chunks of particles. Other examples can be found in coral growth, the path taken by lightning, coalescing of dust or smoke particles, and the growth of some crystals.

2.2 Significance of fractals

The term fractals have always been associated with the complex geometric shapes which can be characterized by non-integer dimensions. Generally, fractals can be found in unbalanced phenomena either naturally or experimentally developed in laboratories. The fractal concept has been used in many fields like chemistry, biology, medicine, weather forecast and engineering where it provides understanding of the extraordinary patterns and chaos (Radnoczy et al., 1987; Chandra & Chandra 1996; Neimeyer et al., 1984).

2.2.1 Fractals in physical sciences

Fractals obviously generate some convincing models of natural phenomena such as mountains and clouds for use in computer graphics imagery, and they provide very compelling abstract pictures. But since 1980's until 1990's, about one third of all physics papers submitted to journals for publication at least mentioned fractals somewhere (Musgrave, 1993). It is also known that many universities all around the world have now offered courses on the subject fractals mainly concerning the field of mathematics and physics.

Looking at fractals in mathematics, some fractal patterns exist only in mathematical theory, but others provide useful models for the irregular yet patterned shapes found in nature such as the branching of rivers and trees. Mathematicians tend to rank fractal dimensions on a series of scales between 0 and 3. One-dimensional fractals (such as a segmented line) typically rank between 0.1 and 0.9, two-dimensional fractals (such as a shadow thrown by a cloud) between 1.1 and 1.9, and three-dimensional fractals (such as a mountain) between 2.1 and 2.9. Most natural objects, when analyzed in two dimensions, rank between 1.2 and 1.6 (Ouellette, 2001).

The nonlinear mathematics models nature more accurately, but is intractable in comparison to the linear approximations. When computers made it possible for scientists to begin to cope with these previously-intractable nonlinear systems, they discovered something very surprising which is in any perturbation to the initial state of the system, no matter how small or seemingly insignificant, will cause the system to diverge; that is to evolve into an arbitrarily different future state, within a finite period of time. This discovery is known as *deterministic chaos* or *sensitivity to initial conditions.*

2.2.2 Fractals in biological sciences

Biologists have traditionally modeled nature using Euclidean representations of natural objects or series. Examples include the representation of heart rates as sine waves, conifer trees as cones, animal habitats as simple areas, and cell membranes as curves or simple surfaces. However, scientists have come to recognize that many natural constructs are better characterized using fractal geometry. Biological systems and processes are typically characterized by many levels of substructure, with the same general pattern being repeated in an ever-decreasing cascade. Relationships that depend on scale have profound implications in human physiology (West & Goldberger, 1987), ecology (Loehle, 1983; Wiens, 1989), and many other sub-disciplines of biology. The importance of fractal scaling has been recognized at virtually every level of biological organization.

Fractal geometry may prove to be a unifying theme in biology (Kenkel & Walker, 1993) since it permits generalization of the fundamental concepts of dimension and length measurement. Most biological processes and structures are non-Euclidean, displaying discontinuities, jaggedness and fragmentation. Classical measurement and scaling methods such as Euclidean geometry, calculus and the Fourier transform assume continuity and smoothness. However, it is important to recognize that while Euclidean geometry is not realized in nature, neither is strict mathematical fractal geometry. Specifically, there is a lower limit to self-similarity in most biological systems, and nature adds an element of randomness to its fractal structures. Nonetheless, fractal geometry is far closer to nature than is Euclidean geometry (Deering & West, 1992).

The relevance of fractal theory to biological problems is dependent on objectives. To the forester interested in estimating stand board-feet, a Euclidean representation of a tree trunk (as a cylinder or elongated cone) may be quite adequate. However, for an ecologist interested in modeling habitat availability on tree trunks (say, for small epiphytes or invertebrates), fractal geometry is more appropriate. Using the approach of fractal geometry, the complex surface of tree bark is readily quantified.

A forester's diameter tape ignores the surface roughness of the bark, giving but a crude estimate of the circumference of the trunk. For an insect 10 mm in length, the distance that it must travel to circumnavigate the trunk is much greater than the measured diameter value. For an insect of length 1 mm, the distance traveled is even greater. This has consequences on the way that the tree trunk is perceived by organisms of different sizes. If the bark has a fractal dimension of $D = 1.4$, an insect an order of magnitude smaller than another perceives a length increase of $10^{D-1} = 10^{0.4} = 2.51$, or a habitat surface area increase of $2.51^2 = 6.31$. By contrast, for a smooth Euclidean surface, $D = 1$ and both insects perceive the same 'amount' of habitat. The higher the fractal dimension D, the greater the perceived rate of increase in length (or surface) with decreasing scale.

2.3 Fractal growth models

Many fractal growth phenomena found in experiments and numerical simulations explored the properties of aggregation kinetics, gelation, and sedimentation (Aharony, 1991). The aggregation of particles often produces fractal clusters. A typical aggregate is the commonly known computer generated simulation of 'diffusion limited aggregation'. The shape looks very similar to those arise in many natural aggregation processes, including diffusion limited electrodeposition (Matsushita et al., 1984), growth in aqueous solutions (Sawada et al., 1986), dielectric breakdown (Niemeyer et al., 1984), viscous fingers in porous media (Maloy et al., 1985), and fungi and bacterial growth (Matsuura & Miyazima, 1992; Matsuyama et al., 1993; Ben-Jacob et al., 1994).

To describe these aggregates, one must first characterize their structures quantitively (Aharony, 1991). Characterization on its fractal dimensionality, or exponents, each of which determines one of its physical properties is very important. Growth models are used to understand the relationship between the microscopic interactions which are responsible for its growth, and the specific complex macroscopic shapes. This is done by setting up a few simple microscopic growth rules, by which particles are added to the aggregate and with repeated iteration it gives rise to the macroscopic cluster. Some of the well known fractal growth models normally used for simulation of fractals are described in the following sections.

2.3.1 Eden model

The Eden model is the simplest growth model (Eden, 1961) and the one that probably applies in most cases. Starting from an initial seed, a new particle is added to cluster on one of the surface sites. A surface site here is defined as a site sharing a side with the existing cluster. The way in which the surface site is chosen can vary. One version of the Eden model selects with equal probability among all the surface sites where a new particle will be added. Another version counts the number of neighbors of each surface site and the probability that a new particle is added is directly proportional to the number of neighbors. The third version of the Eden model chooses a 'mother cell' with equal probability among the particles which are not completely surrounded by other particles.

2.3.2 Percolation model

The randomness of a fluid spreading through a medium maybe of two quite different types (Feder, 1988). The first type is the random walks of the fluid particles in the familiar diffusion processes. The other case in which the randomness is frozen into the medium itself and it is known as a 'percolation process', since it behaves like coffee in a percolator (Broadbent & Hammersley, 1957).
Compared to diffusion process where a diffusing particle may reach any position in the medium, percolation process has a feature, where there exists a 'percolation threshold', under which the spreading process is confined to a 'finite' region. For example, spreads of blight from one tree to the other in an orchard where the trees are planted on the intersections of a square lattice. Here, when the spacing between the trees is increased so that the probability for infecting a neighboring tree falls below a critical value, then the blight will not spread over the orchard. Thus, the value of the percolation threshold has to be determined by simulations.

2.3.3 Ballistic deposition model

Ballistic deposition was introduced as a model of colloidal aggregates, and early studies concentrated on the properties of the porous aggregate produced by the model (Family, 1990: Horvath et al., 1991). The particles in the ballistic deposition model follow a straight-line trajectory until they first encounter a particle on the surface, or a particle in one of the nearest-neighbor columns. As soon as a particle reached such a position, it permanently sticks to the surface and becomes part of the deposit. Evolution of an interface in a ballistic deposition model can be described by the dynamic scaling approach (Family & Vicsek, 1985). Moreover the surface of the deposit is a self-affine fractal, since the atoms are not allowed to diffuse on the surface.

2.3.4 Dielectric breakdown model

Dielectric breakdown refers to the formation of electrically conducting regions in an insulating material exposed to a strong electric field. For example, the intense electric fields during thunderstorms can produce a conducting path in the air along which many electrons flow (lightning). A formal model, ignoring the physical details of the processes, was proposed in 1984 by Niemeyer, Pietronero and Weismann (Niemeyer et al., 1984). Dielectric breakdown patterns exhibit a branching, fractal pattern with a dimension of about 1.7.

2.3.5 Viscous fingering model

In viscous fingering the principal force is due to viscous forces in the defending fluid (Aharony, 1991). The process is obtained by injecting a low viscosity fluid into a medium of high viscosity fluid with a high injection rate. The capillary effects and the pressure drop in the invading fluid are negligible. The structures typically consist of fingers of invading fluid that propagate through the medium with only a few small trapped clusters of defending fluid left behind. Viscous fingering was first studied in a Hele-Shaw channel where one observes fingering patterns when glycerol is displaced by air (Saffman & Taylor, 1958). A Hele-Shaw cell consists of two transparent plates separated by a given distance and the patterns obtained are fully described by Darcy's equation and the capillary pressure due to the interfaces between the two phases. In 1985, Chen and Wilkinson (Chen & Wilkinson, 1985) and Måløy and coworkers (Måløy et al., 1985) studied viscous fingering in a porous medium where they concluded that the disorder of the system has significant effect on the fingering process.

2.3.6 Diffusion limited aggregation model

Diffusion limited aggregation (DLA) is a model of irreversible growth to generate fractal structures as proposed by Witten and Sander (1981). It has been used to study a great variety of processes including dendritic growth, viscous fingering in fluids, dielectric breakdown and electrochemical deposition. The model is set by the following simple rules:

> A seed is fixed at the origin of some coordinate system and one particle is released from a far-away boundary and allowed to take random walks (diffuse). If the particle touches the seed, it irreversibly sticks to the seed and forms a two-particle aggregate. As soon as the random walker is removed either by being captured or escaping the boundary, the next walker is released and the process is repeated. Now it can stick to any particle in the aggregate as well as the original seed.

The resulting clusters are highly branched since DLA enhances the instability of growth. The arriving particles are far more likely to stick to the tips of outer branches than to maneuver their way deep into the *fjords* (narrow inlet of a section) before contacting the surrounding branches. Thus the tall branches of the cluster screen the small ones and grow faster. The growth on the tips, however, is not always in the outward radial direction. Sometimes a few new branches are spun off from one tip site as occurred in the original seed. The *tip-splitting* makes the DLA clusters a self similar fractal.

2.4 Methods for determination of fractal dimension

Fractal dimension is a statistical quantity that gives an indication of how completely a fractal appears to fill space, as one zooms down to finer and finer scales. There are many specific definitions of fractal dimension. Summary of some of the more commonly used methods for determination of fractal dimension of natural forms are presented in this section. These methods include the information dimension method, mass dimension method and box counting method.

i. Information dimension method

This method requires the use of boxes but is generally different from the box counting method. It does not consider the number of boxes occupied regardless of whether it contains one point of relatively large number of points. Instead, the information dimension effectively assign weights to the boxes in such a way that boxes containing a greater number of points count more than boxes with fewer points (Dierking, 2000). The fractal dimension D_i is given from the proportionality

$$I(d) \sim -D_i \log(d) \tag{5}$$

with $I(d)$ is the information entropy of $N(d)$ boxes of size d, given by

$$I(d) = -\sum_{i=1}^{N(d)} m_i \log(m_i) \tag{6}$$

with $m_i = \frac{M_i}{M}$ where M_i is the number of points in the ith box and M the number of total points in the data set.

ii. Mass dimension method

The mass dimension method also known as the Scholl method, which yields the fractal dimension D_m, following the proportionality

$$m(r) \sim r^{Dm} \tag{7}$$

where $m(r) = M(r)/M$ is the 'mass' within a circle of radius r, where $M(r)$ is the data set of points contained within a circle and M the total number of points in the set. If the set is a fractal, the plot of log $m(r)$ versus log r will follow a straight line with a positive slope equal to D_m (Dierking, 2000). This method is best suited to objects that follow some radial symmetry, such as the dendritic growth in radial axis.

iii. Box-counting dimension

In fractal geometry, the box-counting dimension is a way of determining the fractal dimension of a set S in a Euclidean space R^n. To calculate this dimension for a fractal S, imagine this fractal lying on an evenly-spaced grid, and count how many boxes are required to cover the set. The box-counting dimension is calculated by seeing how this number changes as the grid becomes finer.

Suppose that $N(s)$ is the number of boxes of side length s required to cover the set (Hastings & Sugihara, 1993), then S has box dimension D if $N(s)$ satisfies the power law

$$N(s) \approx c(1/s)^D \tag{8}$$

asymptotically in the sense that

$$\lim_{s \to 0} N(s) s^D = c \tag{9}$$

By solving equation (8) asymptotically for D, the box-counting dimension is computed as:

$$D = \lim_{s \to 0} \left[-\frac{\log N(s)}{\log s} \right] \tag{10}$$

This method is a favorite among most researchers and is considered the easiest to perform (McNamee, 1991). The box-counting dimension can be used to analyze irregularities in surfaces filling space volume and suitable for images, however complex. The use of a mesh grid overlapped over a structure allows the box counting method to conduct both textural and structural analysis of a structure. In addition, the mesh grid also allows the analysis of objects scattered in an image and this method can be adapted to measure objects or processes in multiple dimensions (Cross, 1997).

3. Simulation of fractals

3.1 Simulation of fractals using DLA model

DLA is one of the most important models of fractal growth. This model is based on the Brownian motion theory. It refers to a simple growth algorithm in which individual particles are added to a growing cluster through diffusion-like process. Starting from any suitable immobile aggregate seed in a plane, a new particle is launched at a random position far away from the aggregate seed and is allowed to undergo Brownian motion. When the random walking particle touches the seed, it is stopped and incorporated to the aggregate.

The process of launching a random walker and adding it to the aggregate on its first contact is repeated until the aggregate reached a desired number of particles (Witten & Sander, 1981). Figure 4 gives a visual representation of the above mentioned process.

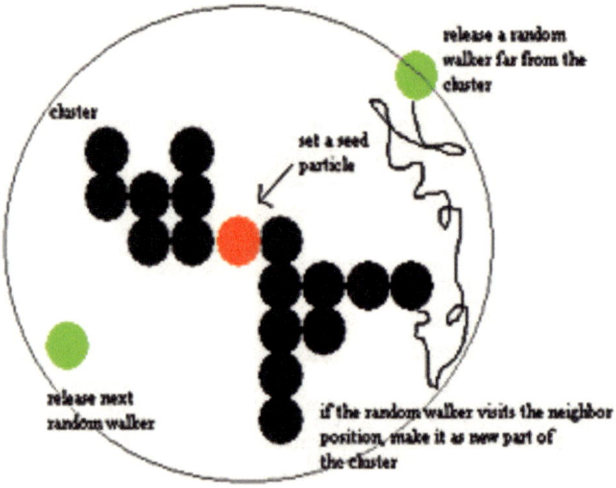

Fig. 4. An off the scale model of aggregation of cluster particles

3.2 Simulation of fractal pattern using a grammar based model (L-systems)

L-systems; a mathematical formalism as a foundation for an axiomatic theory of biological development (Lindenmayer, 1968) was proposed by a biologist, Aristid Lindenmayer in 1968. L-systems have found several applications in computer graphics especially in areas which include generation of fractals and realistic modeling of plants. Central to L-systems, is the notion of rewriting, where the basic idea is to define complex objects by successively replacing parts of a simple object using a set of rewriting rules or productions. The rewriting can be carried out recursively.

Aristid Lindenmayer's L-systems introduced a new type of string rewriting mechanism. In L-systems, grammars productions are applied in parallel, replacing simultaneously all letters in a given word. There are a number of different types of L-systems. The two major classifications are reflected in the naming conventions:

1. Deterministic
2. Stochastic

The two classes of L-systems make it possible to generate simple and complex geometric patterns in the study of fractals.

3.2.1 Deterministic L-systems

Also known as D0L-systems, deterministic L-system is the simplest classes of L-systems. D0L stands for deterministic and 0-context or context-free L-systems. The rewriting process starts from a distinguished string or initiator called the axiom, and followed by a generator (rules) that is applied to the axiom to generate a new string. This generation can be iterated to produce strings of arbitrary length, which then can be interpreted as a series of turtle commands by a turtle graphic system (Abelson & diSessa, 1982). This turtle concept is explained in detail in section 3.2.3.

3.2.2 Stochastic L-systems

L-systems are usually deterministic and they provide description based on individual patterns. Consequently, every time a word is derived using a given system, the resulting words will be the same (Prusinkiewicz & Lindenmayer, 1990). This may lead to restrict a certain pattern to a certain form and would not always be desired. Imagine visualizing complex fractal patterns found in many natural processes, such as electrodeposition, growth in aqueous solutions, dielectric breakdown, viscous fingers in porous media, and fungi and bacterial growth. If only one L-System was used to describe every pattern, then the results would look unrealistic (Kaandorp, 1994). On the other hand, creating a single L-System with individual productions for each pattern would be a tedious task, and it still could not guarantee similarity between individuals.

Instead of using only one L-system to describe these patterns, it is wiser to use probabilistic/stochastic L-Systems. Each of these production/rules is assigned a probability. All fractal growth patterns generated by the same deterministic L-system are identical. An attempt to combine them in the same picture would produce a striking, artificial regularity (Prusinkiewicz & Lindenmayer, 1990). In order to prevent this effect, it is necessary to introduce one-to-one variations that will preserve the general aspects of a pattern but will modify its details. Variations can be achieved by randomizing the turtle interpretation, the L-system, or both.

In order to achieve this, it is necessary to use suitable production rules with the implementation of the turtle graphics command in a turtle graphics system. Stochastic L-Systems was considered a more suitable simulation technique for the simulation of fractals formed without using any external stimuli.

3.2.3 Graphical representation of L-systems

The most common turtle interpretation used in L-system today is based on the LOGO-style turtle (Abelson & diSessa, 1982), as introduced by Prusinkiewicz (1986). The main concept is that some modules in the L-system string are interpreted as commands executed by a turtle. In 2D, the *state of the turtle* (*S*) is characterized by its position and orientation. Turtle graphic interpretations can exhibit different levels of complexity (Alfonseca & Ortega, 2001). Papert (Papert, 1980) created turtle graphics in 1980, describing it as a trail left by an invisible 'turtle' whose state at every instant is defined by its position and the direction in which it is looking. Set of instructions (commands) to the turtle (Peitgen et al., 1992) are explained as follows:

- F moves the turtle one step forward, in the direction of its current angle, leaving a visible trail. We call F a 'draw 'letter.
- *f* moves the turtle one step forward, in the direction of its current angle, with no visible trail.
- +(plus) increases the turtle angle by θ.
- -(minus) decreases the turtle angle by θ.
- [stacks the current position and orientation of the turtle.
-] moves the turtle invisibly to the position and orientation stacked at the top of the stack and pops it.

A state of the graphic turtle is defined as a triplet (x, y, θ), where the Cartesian coordinates (x, y) represent the turtle's position in 2D space, and the heading, the angle θ is interpreted as the direction in which the turtle is heading.

The following is an example of an L-system specification.

Axiom: F
Production rules: F → F[+F]F[-F] [F]
Angle, θ:30°

Derivation of a string from an L-system is done in much the same way that a string is formed from a traditional grammar. It begins with the axiom, F. At any point in the derivation, all Fs in a string are replaced in parallel by a particular replacement string. In the L-system described above, each F in a string is replaced by F[+F]F[-F] [F] . So, starting with the axiom, F, the string after one substitution would be

F[+F]F[-F] [F]

and the string after two substitutions would become

F[+F]F[-F] [F] [+F[+F]F[-F] [F]]F[+F]F[-F] [F] [-F[+F]F[-F] [F]][F[+F]F[-F][F]]

A generated object is often referred to by the number of string substitutions that has been performed. The numbers of iterations are defined as being one greater than the number of string substitutions. Figure 5 illustrates the objects for four different iterations that have been generated from the L-system described above.

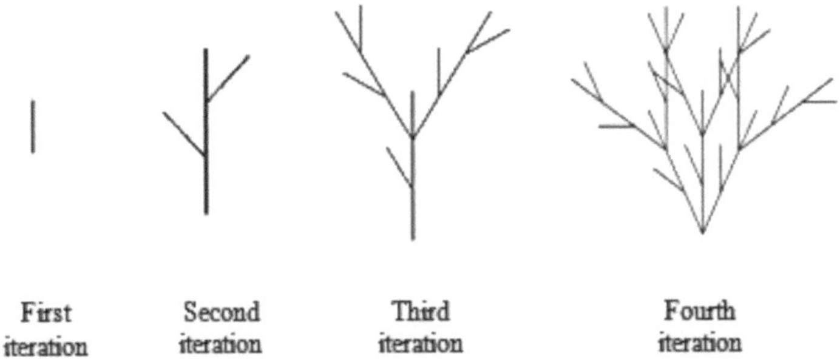

Fig. 5. Objects generated from an L-system until the fourth iteration

4. Fractals in ion conductive polymer electrolytes

Eversince the introduction of the 'Fractal Geometry' concept by Mandelbrot (Mandelbrot, 1983) in 1977, much work was concentrated on theoretical simulation/modeling of this concept. Theoretical simulation of fractal patterns, in particular Diffusion Limited Aggregate (DLA) requires particles of uniform as well as non-uniform size performing random walk. In the case of polymer electrolyte membranes, studies have been done to develop polymer electrolytes with high ionic conductivities especially in the field involving advanced materials known as superionic solids or fast ionic conductors (Amir et al., 2010b). In this type of conductors, the conductivity is due to the motion of ions. Superionic solids or fast ionic conductors brought about the development of high energy density batteries (Armand et. al, 1979; Murata, 1995; Borghini et. al, 1996), electrochomic devices (Ratner, 1987; Scrosati, 1990), fuel cells (Prater, 1990), chemical sensors (Somov et. al, 2000) and capacitors (Pernaut and Goulart, 1996) etc. Initially, the fractals observed in the polymer electrolytes were discovered by chance. Some of the fractals observed in the polymer electrolyte membranes are shown in Figures 6(a)-(c). The study of these fractals may be useful in understanding the movement of ions in the polymer electrolyte membranes.

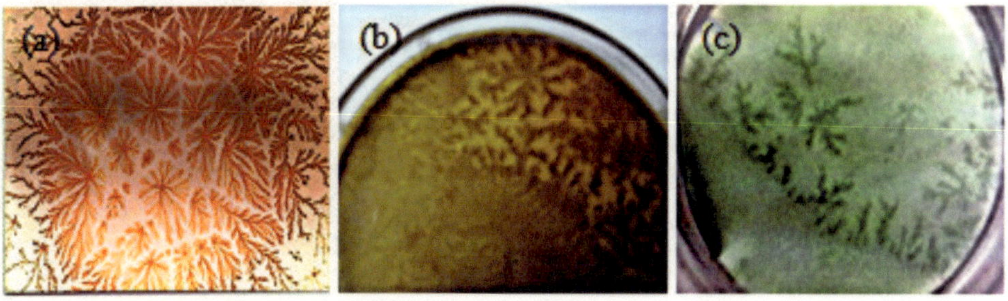

Fig. 6. Fractal aggregates of different sizes in (a) chitosan (b) PEO and (c) PVDF-HFP based electrolyte membranes

4.1 Fractal growth patterns identification and simulation

It has been identified that the formation of fractals without using any external stimuli resulted into isotropic DLA patterns as reported by Chandra (1996) and Amir et al. (2010a; 2011b). Amir et al. (2010a; 2011a; 2011b) have succeeded to obtain fractal aggregates of different sizes in the films such as PEO, chitosan and PVDF-HFP polymers infused with an inorganic salt without any external stimuli. As can be observed in Figure 6, fractals are formed at the different nucleation centers and then grow in certain directions away from the nucleation site. The fractals grow irregularly and in an unpredictable motion. Among the techniques used to simulate such fractal patterns are DLA model which is based on the Brownian motion theory and fractal dialect called L-systems.

4.2 Simulation of fractals in ion conductive polymer membranes

Simulation of the DLA model gives a simple yet effective way to represent fractals obtained in polymer membranes. On the other hand, the simulation using the L-system technique provides a general approach on how to visualize growth of fractals with respect to the production rules involved and the governing number of iterations required to actually simulate a model that best represent the original pattern observed in the experimental outcome.

4.2.1 Simulation using DLA model on a square lattice

Computer simulations of the fractals are performed based on the DLA model described earlier. The simulation starts with a single seed at the centre of a square lattice. Then the seed will start to grow gradually until a full single cluster is formed. This cluster grows outward, one generation after another. The basic algorithm of the whole process is as follows:

1. A list called **occupiedSites** is created, containing the lattice site {0, 0}.
2. Determine the lattice site nearest to a randomly chosen location along the circumference of a circle whose radius, rad, equals a specified value, s, plus the maximum absolute coordinate value in **occupiedSites**.
3. Starting at the selected lattice site, execute a lattice walk until the step location is either at a distance greater than (rad + s), or on a site that is contiguous (adjacent) to a site in **occupiedSites**. Call the final step location of the walk, **loc**.
4. Check if **loc** is adjacent to a site in the occupiedSites list and if it is, add loc to occupiedSites.
5. Execute the sequence of steps 2 through 4 until the length of **occupiedSites** reaches a value n.

Figure 7 gives an illustration of the implementation of the DLA algorithm.

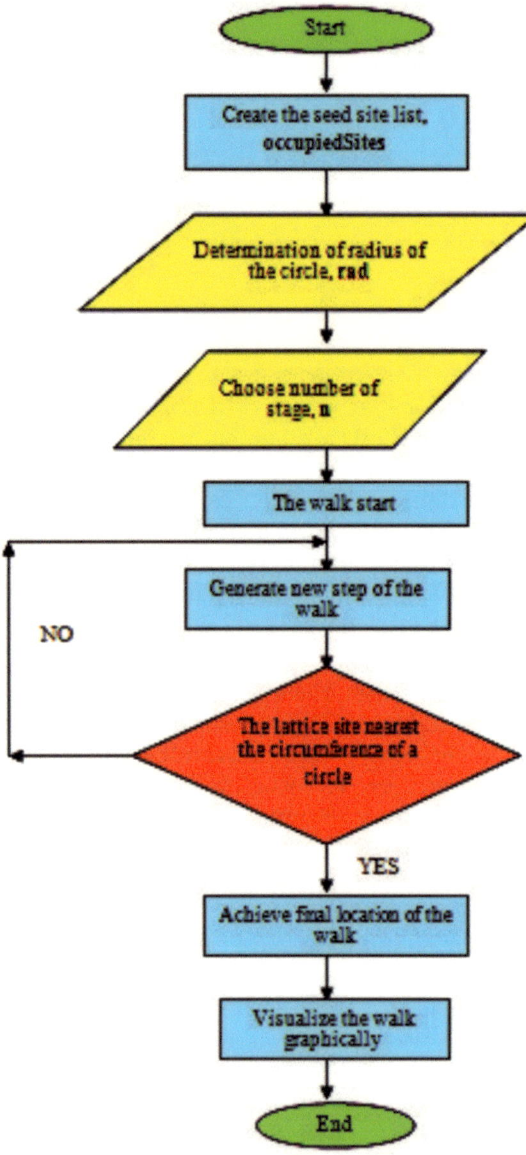

Fig. 7. An illustration of the simulation of DLA model on a square lattice

In summary, the algorithm adds a particle as a result of random walk of a drifter from the perimeter to a point on the boundary of the structure. The drifters were introduced into the grid at a point chosen at random from the points equidistant from the origin, where the initial particle was placed. If the movement of a drifter would take it beyond this perimeter, it is reflected back to the interior. If it moves to a point on the boundary of the structure, it is transformed into a part of the structure. This process is repeated until the desired stage has been achieved.

For simulation purposes, the images of the fractal patterns such as those displayed in Figure 8 are chosen specifically.

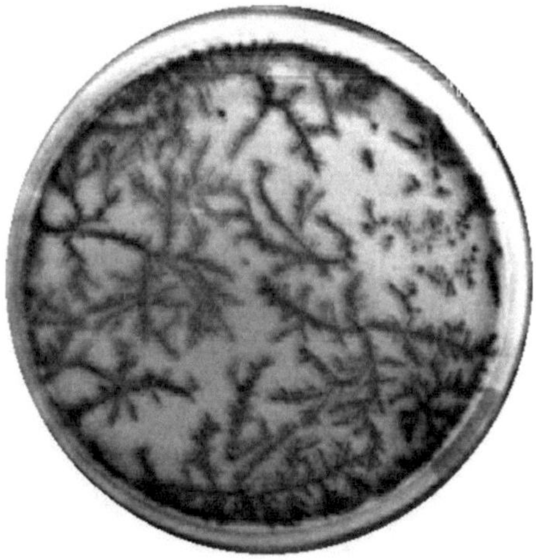

Fig. 8. DLA fractals in a PEO-NH_4I membrane

The simulation of the fractal patterns can be carried out on a square lattice. The animation for a single cluster of the pattern can be created using a computer program developed in order to show how the cluster grows starting from the original nucleation centers that eventually grow according to a specific size desired. There are some properties of the clusters that have been identified. These properties are as follows:

1. Branching and screening
   The random growth process leads to the formation of small tips which are likely to capture diffusing particles. They *screen* their surroundings which later have the effect of screening that will *self-stabilize* the tip until it grows even larger forming new tips which are delicate, branched and tree-like objects.
2. Scale invariance, lack of a typical length-scale
   As the stage of growth increases, it seems like they are a hierarchy of arms, branches, twigs and sprouts with *fjord* like empty regions of all sizes.
3. Stochastic self-similarity
   Substructures of the cluster are found to be self-similar i.e. they can be reproduced after proper rescaling.

Apart from the identified properties of the DLA cluster, it is also important to know the growth site probabilities for every particle released during the simulation. It has been discovered that diffusing particles are highly unlikely to wander into one of the inner *fjords.* The diffusing particles have a high probability to attach to the protruding tips. They are already visible from the marking of most recent particles and growth occurs essentially only in a small *active zone* within the predetermined radius.

Figures 9(a) (i-v) present the fractals of PEO-NH_4I (60:40 wt %) membrane as observed in figure 8. Their patterns simulated using DLA model are shown in Figures 9(b) (i-v).

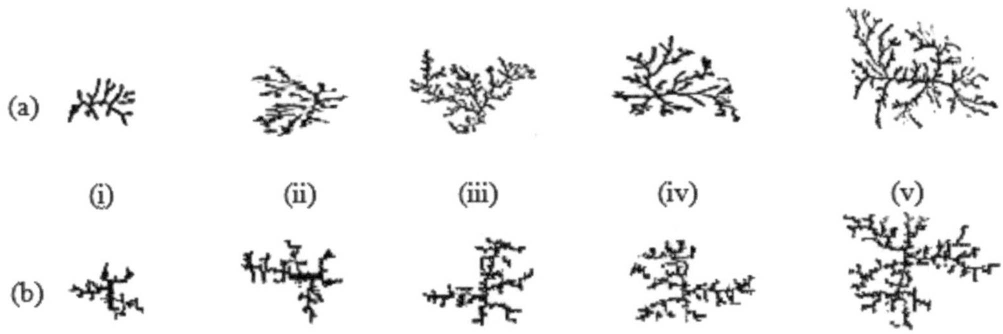

Fig. 9. The experimentally observed (a) and simulated (b) DLA patterns in PEO-NH_4I system using DLA model

The fractal dimensions for the simulated and cultured fractals are listed in Table 2. The fractal dimensions of the cultured fractals are determined using a computer software tool, utilizing the box-count method, developed by Suki et. al. (2007) while those of simulated fractals are calculated automatically by a subprogram incorporated in the simulation program. The table shows that the fractal dimension values of the simulated fractals are comparable with the fractal dimension values obtained from their respective experimentally cultured ones. For example, the fractal dimension of experimentally cultured fractal in Figure 9(a) (v) is 1.742 ± 0.044 while its corresponding simulated fractal, Figure 9(b) (v) has a fractal dimension of 1.778 ± 0.043. In every calculation of the fractal dimension, the error values for both experimentally and simulated patterns are found to be so small with standard deviation of less than 1.6%.

Experimentally cultured fractals		Simulated fractals using DLA model		Percentage of difference (%)
Figure	Fractal dimension	Figure	Fractal dimension	
9(a)(i)	1.688 ± 0.049	9(b)(i)	1.725 ± 0.044	1.29
9(a)(ii)	1.719 ± 0.043	9(b)(ii)	1.765 ± 0.049	1.56
9(a)(iii)	1.754 ± 0.047	9(b)(iii)	1.759 ± 0.042	0.16
9(a)(iv)	1.706 ± 0.045	9(b)(iv)	1.747 ± 0.045	1.41
9(a)(v)	1.742 ± 0.044	9(b)(v)	1.778 ± 0.043	1.19

Table 2. Fractal dimension values of the fractals simulated using DLA model and their respective experimentally cultured fractal patterns.

From the table, it is clear that the percentage difference of fractal dimension values between the experimental and simulated patterns are marginally close. The highest percentage difference is found to be less than 1.6% that is 1.56%. The lowest percentage difference is 0.16%. These show that the simulated fractal patterns were of fairly good conformity with the fractal patterns observed in the PEO-NH_4I polymer films.

The small difference of fractal dimension values between the experimental and simulated patterns maybe attributed to the way the simulation was done in the simulation. The simulation of fractal patterns using the modeling of DLA on a square lattice written in a computer program chosen for this study has a run time restriction. To actually simulate a larger fractal pattern requires a longer time to complete as a particle that moves close to the cluster has to investigate all neighbor sites, whether these already belong to the cluster. The particle should either stick or walk freely. The information about the neighborhood should be assigned to each site, so that a walker only make contact with the site which it is on instead of all four (in a square lattice) possible neighbors.

4.2.2 Fractal growth simulation using L-systems

Stochastic L-system allows various shapes to be drawn. The recursive nature of the L-system rules leads to self-similarity and thereby complex geometric patterns which are easy to describe with an L-system. The rules of the L-system grammar are applied iteratively starting from the initial state which is called the axiom and a set of production rules.

To develop an L-system for a particular simulation of the complex geometric patterns, some steps (Hashim_Ali et al., 2000) have to be followed:

i. the fractals must be analyzed to infer its stages of growth.
ii. define the axiom and production rules into a string of symbols that assigned a particular meaning.
iii. execute the rules as a computer program and show the results as a graphical output and calculate its fractal dimension.
iv. compare the simulation with the real patterns obtained in the polymer membranes.

As observed in Figure 8, the fractal patterns found in polymer electrolytes take on a branching structure. The branching structure is represented by the square bracket symbols (refer section 3.2.3). When a branch point is reached, the turtle encounters the left square bracket '[' where it should remember its current position and heading. This is called the state of the turtle. Technically the state S is given by $S = \{x, y, \theta\}$. In mathematical terms, the turtle has a state consisting of its current position, given by two coordinates x and y, and a current heading, specified by an angle θ. On the other hand when the turtle reaches the corresponding closing bracket ']' the commands are terminated and the turtle will then return to the branching point which it must remember. In the analysis of fractal growth patterns, there are a few important factors to be considered in performing the simulation. The main criterion is the structure of the fractal such as the number of branches, the size of the structure, and most of all the production rules that suit the original pattern. For the fractals shown in Figure 8, to achieve a good result, three key components in the simulation were identified. They are as follows:

i. Generally, the numbers of branches in most of the observed fractal patterns are about four.
ii. The size of the object differs from one to the other but to obtain the best model for the structure; when running the simulation, the stage of iteration should be at least six. The purpose of choosing the stage of iteration is because if the stage of iteration was too low or too high, it would be hard to get a generally satisfying model for the structure. This is to ensure a 'mature' fractal growth pattern is obtained that resembles the real fractals.
iii. To best describe the growth of complex geometric patterns in polymer membrane, three production rules have been chosen and they are: F[+FF]F[-FF]F, F[+FF]F, and F[-FF]F.

Figures 10(a) (i-v) present the fractals observed in the PEO-NH_4I membrane shown in Figure 8. Their patterns simulated using L-systems are shown in Figures 10(b) (i-v).

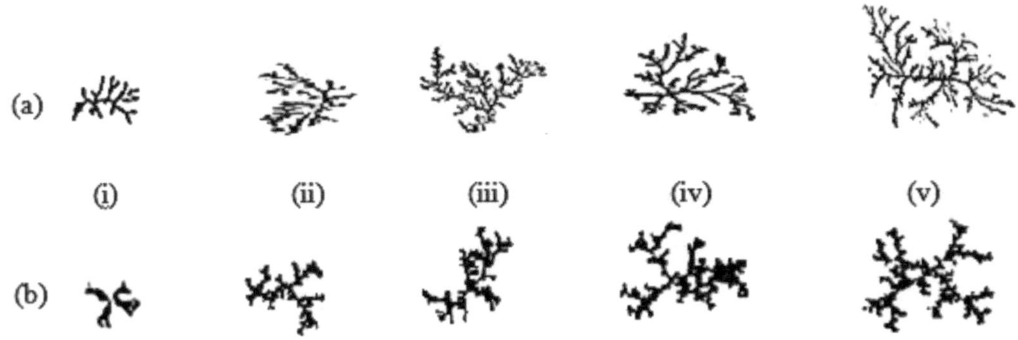

Fig. 10. The experimentally observed (a) and simulated (b) DLA patterns in PEO-NH_4I system using L-Systems

Table 3 gives the fractal dimensions for the original and simulated fractals. The table shows that the fractal dimension values of the simulated fractals are comparable with the fractal dimension values obtained from their respective experimentally cultured ones. For example, the fractal dimension of experimentally cultured fractal in Figure 10(a) (v) is 1.742 ± 0.044 while its corresponding simulated fractal, Figure 10(b) (v) has a fractal dimension of 1.751 ± 0.045. The error values for both experimentally and simulated patterns are found to be so small with standard deviation of less than 1.6%.

Experimentally cultured fractals		Simulated fractals using L-Systems		Percentage of difference (%)
Figure	Fractal dimension	Figure	Fractal dimension	
10(a)(i)	1.688 ± 0.049	10(b)(i)	1.645 ± 0.052	1.51
10(a)(ii)	1.719 ± 0.043	10(b)(ii)	1.756 ± 0.043	1.25
10(a)(iii)	1.754 ± 0.047	10(b)(iii)	1.737 ± 0.047	0.55
10(a)(iv)	1.706 ± 0.045	10(b)(iv)	1.761 ± 0.044	1.54
10(a)(v)	1.742 ± 0.044	10(b)(v)	1.751 ± 0.045	0.26

Table 3. Fractal dimension values of the fractals simulated using L-system and their respective experimentally cultured fractal patterns.

In the simulation using L-system technique, the branches grew completely one after the other from their nucleation center and thus making it difficult to get a complete full grown cluster that matched exactly as the experimentally cultured fractal patterns. These are among the factors which make it difficult to get absolute accuracy thus giving way to the small percentage differences of fractal dimension values between the experimental and simulated patterns.

5. Future research

Fujii et al. (1991) have studied the growth of fractal patterns in a conducting polymer. Furthermore, studies done by Shui et al. (2004) and Rosso (2007) have also gained significant improvements toward the understanding of these phenomena. However the effects of such phenomena in secondary battery have not been fully understood. It is difficult to actually study directly the growth of fractal pattern that forms in the electrode since the fractal patterns could be easily damaged during accumulation. Thus as a substitute, fractals can be cultured in ion conducting polymer electrolyte membrane to replicate the condition in a similar environment via laboratory experiments. With this simple approach, study of the temporal images of the fractal growth pattern by utilizing a programmable image data acquisition device can also be done to get a more accurate simulation. For future research, extension of the basic DLA model or modifications on the L-systems can be carried out to get better results for the morphological evaluations of the fractal growth patterns. Then the dependence of the fractal dimension on the stages of growth can be evaluated and the effect of fractal dimension on the growth process in laboratory scale can further be investigated.

6. Conclusion

Many studies on fractals have been carried out either in applications, usually involving experimental works, or in theory, where most simulations on fractal patterns models are on nature-based fractals such as river flows, coastline and tree branching. This chapter focuses on the simulation of laboratory cultured fractals using ion conductive polymer electrolyte membranes as the media of growth. The simulation developed here is a DLA model based on the Brownian motion theory and a fractal dialect known as L-systems. A computer program has been developed to simulate and visualize the fractal growth. This computer program was also built to calculate the fractal dimension values of the simulated fractal patterns. Comparisons of the fractal dimension values between the laboratory cultured and the simulated fractals indicate an acceptable conformity with each other.

7. Acknowledgment

The authors would like to express gratitude and sincere appreciation to all the individuals whom directly and/ or indirectly helped realized the success of this research work. Financial support given by the University of Malaya, in terms of Short Term Research Funds (FP007/2005D and FS123/2008B) is greatly acknowledged.

8. References

Abelson, H. & diSessa, A. A., (1982) *Turtle geometry*. M.I.T. Press, Cambridge.

Addison, Paul S., (1997) *Fractals and Chaos - An Illustrated Course*. Institute of Physics (IoP) Publishing

Aharony, A., (1991), in: *Fractals and Disordered Systems,* ed. by Bunde, A. & Havlin, S.: Springer, p.177-178.

Alfonseca, M. & Ortega, A. (2001)*Determination of fractal dimensions from equivalent L systems* IBM J. RES. & DEV. 45, 6, 797-850.

Amir, S., Mohamed, N. S. & Hashim Ali, S. A. (2010a) *Cent. Eur. J. Phys.* Vol. 8(1) pp. 150-156

Amir, S., Mohamed, N. S., Subban, R. H. Y. (2010b) 'The Investigation on Ionic Conduction of PEMA Based Solid Polymer Electrolytes', *Advanced Materials Research* 93 - 94 381-384

Amir, S., Mohamed, N.S. & Hashim Ali, S. A. (2010c) Using Polymer Electrolyte Membranes as Media to Culture Fractals: A Simulation Study, *Advance Materials Research,* 93-94 35-38

Amir, S, Hashim Ali, S. A. & Mohamed, N. S. (2011a) Studies of fractal growth patterns in poly (ethylene oxide) and chitosan membranes. *Ionics*. 17(2) 121-125.

Amir, S. Hashim Ali, S. A. & Mohamed, N. S. (2011b). Ion Conductive Polymer Electrolyte Membranes and Simulation of Their Fractal Growth Patterns. *Sains Malaysiana* 40(1) 75-78

Armand, M.B, Chabagno, J.M. and Duclot, M.J. (1979), in: Fast Ion Transport in Solids, (eds.) Vashita, P., Mundy, J.N. and Shenoy, G.K., Elseiver North-Holland, New York, 131

Barkey, D., (1991). *J. Electrochem. Soc.* 138: 2912.

Ben-Jacob, E., Shochet, O., Tenenbaum, A., Cohen, I., Czirok, A. & Vicsek, T. (1994). *Nature* 368, 46.

Borghini, M.C., Mastragostino, M. and Zanelli, A. (1996), 'Investigation on lithium/polymer electrolyte interface for high performance lithium rechargeable batteries', *J. of Power Sources* 68 52-58

Broadbent, S. R. & Hammersley, J.M. (1957). Percolation Process I. Crystals and Mazes. *Proc. Cambridge Philos. Soc.* 53, 629-641.

Carr, B.J. & Coley, A.A. (2003). Self-similarity in general relativity, Available from http://users.math.uni-potsdam.de/~oeitner/QUELLEN/ZUMCHAOS/selfsim1.htm

Chandra, A. & Chandra, S. (1994) *Physical Review B. The American Physical Society* 49, 1, pp. 633-636

Chandra, A. (1996) *Solid State Ionics* 86-88 , pp. 1437-1442

Chandra, A. & Chandra, S. (1993). *Current Sci.* 64, pp. 755.

Chen, J.-D. & Wilkinson, D. (1985). Pore-scale viscous fingering in porous media. *Phys. Rev. Lett.*, pp. 1892-1895.

Combes, F. (1998). Fractal Structures Driven by Self-gravity: Molecular Clouds and the Universe. *Springer* 72, pp. 1-2.

Cross SS. (1997). Fractals in pathology. *J Pathol* 182, pp. 1–8.

Dargahi-Noubary, G. R. (1997). A Test of the Cyclicity of Earthquakes, *Natural Hazards* 16 pp. 127–134.

Deering, W. & West, B.J. (1992). Fractal Physiology. *IEEE Engin. Med. Biol.* 11, pp. 40-46.

Dierking, I. (2000). Fractal Growth of the Liquid Crystalline B2 Phase of a Bent-core Mesogen. *J. Phys:C:Condens. Matter.* 13, pp. 1353-1360.

Ebert, D. S.(1996). Advanced Modeling Techniques for Computer Graphics, *ACM Computing Surveys*, 28(1), pp. 153-156.

Eden, M., (1961). *Proc. 4th Berkeley Symp. on Math. Stat. and Prob.* 4, pp. 223.

Family, F. & Vicsek, T., (1985), *J. Phys. A: Math. Gen.*18, pp. L75.

Family, F., (1990). *Physica A* 168, pp. 561

Feder, J.(1988). *Fractals*, New York: Plenum Press.

Focardi, S.M. (2003). Fat tails, scaling and stable laws: A critical look at modeling extremal events in economic and financial phenomena, Available from http://www.theintertekgroup.com/scaling.pdf

Fujii, M.; Arii, K. & Yoshino, K. (1991). The Growth of Dendrites of Fractal Pattern on a Conducting Polymer. *J. Phys.: Condens. Matter*, 3, pp. 7207-7211

Golubev, Yu. N., Fomin, V. V. & Cherkesov, L. V. (1987). Interaction between surface gravitational waves and local rise of sea-bed in the uniform ocean, Physical Oceanography, *Springer New York*, 1, 1, pp. 3-9.

Hashim Ali, S. A.; Mohamed, N. S.; Shariff, A. A. & Arof, A. K. (2000). Simulating fractal growth in chitosan doped silver nitrate film using L-system, *Solidstate Ionic Devices: Science & Technology*, pp. 16-19

Hastings, H.O. & Sugihara, G.(1993). *Fractals: A Users Guide For The Natural Science*, New York: Oxford University Press.

Horvath, V. K., Family, F. & Vicsek, T., (1991). *Phys. Rev. Lett.* 67, pp. 3207.

Janke, W. & Schakel, A. M. J. (2005). *Phys. Rev. Lett.* 95, pp. 135702

Judd, C.(2003). Fractals - Self-Similarity, 17.06.2007 Available from http://www.bath.ac.uk/~ma0cmj/FractalContents.html

Kaandorp, J.A. (1994). *Fractal modelling: growth and form in biology.* Springer-Verlag, Berlin, New York.

Kaufmann, J. H., Nazzal, A. I. & Melroy, O. R. (1987). *Phys. Rev. B* 35, pp. 1881-1890.

Kenkel, N.C. & D.J. Walker. (1993). Fractals and ecology. *Abst. Bot.* 17,pp. 53-70.

Lindenmayer, A. (1968)."Mathematical models for cellular interaction in development." *J. Theoret. Biology*, 18, pp. 280-315.

Lo Verso, F.; Vink, R. L. C.; Pini, D. & Reatto, L. (2006). *Physical Review E* 73, pp. 061407

Loehle, C. (1983). The fractal dimension and ecology. *Specul. Sci. Tech.* 6, pp. 131-142.

Maloy, K.J., Feder, J. & Jossang, T. (1985). *Phys. Rev. Lett.* 55, pp. 2688.

Mandelbrot, B. B. (1983). *The Fractal Geometry of Nature*, San Francisco: W. H. Freeman and Co.

Mandelbrot, B., (1977). *Fractals: Form, Chance and Dimension,*; W H Freeman and Co.

Marcone, B.; Orlandini, E. & Stella, A. L. (2007). Knot localization in adsorbing polymer rings, *Physical Review E* 76, pp. 051804

Matsushita, M., Sano, M., Hayakawa, Y., Honjo, H. & Sawada, Y. (1984). *Phys. Rev. Lett.* 53, pp. 286.

Matsuura, S. & Miyazima, S. (1992). *Physica A* 191, pp. 30.

Matsuyama, T., Harshey, R. M. & Matshushita, M.(1993). *Fractals* 1, pp. 336.

Murata, K. (1995), 'An overview of the research and development of solid polymer electrolyte batteries', *Electrochim. Acta* 40 2177-2184

McNamee, J. E. (1991) Fractal Perspectives in Pulmonary Physiology. *J. Appl. Physiol.* 71(1), pp 1-8

Meakin, P. (1988),Simple Models for Crack Growth, *Crystal Prop. Prep.17* and*18*, pp. 1-54.

Meakin, P. (1991), Models for Material Failure and Deformation, *Science252*, pp. 226-234.

Musgrave, F. K., (1993). *Methods for Realistic Landscape Imaging*, Yale University

Niemeyer, L., Pietronero, L.& Wiesmann, H.J. (1984). *Phys. Rev. Lett.* 52, pp. 1033 -1036

Okubo, S., Mogi, I., Kido, G. & Nakagawa, Y. (1993). *Fractals* 1, pp. 425.

Ouellette, J.,(2001)Pollock'sFractals, 27.10.2007, Available from http://discovermagazine.com/2001/nov/featpollock

Papert, S.(1980). *Mindstorms: Children, Computers, and Powerful Ideas.* Basic Books, New York,

Peitgen , H.O. & Saupe, D.(1988). *The Science of Fractal Images*, Berlin: Springer-Verlag.

Peitgen , H.O., Jurgens, H. & Saupe, D.(1992). *Fractals for the Classroom*, New York: Springer-Verlag.

Prusinkiewicz P. (1986). *Graphical applications of L-systems.* Proceedings Graphics

Prusinkiewicz, P. & Lindenmayer, A. (1990). *The Algorithmic Beauty of Plants*. Springer-Verlag, New York.
Radnoczy, G., Vicsek, T, Sander, L.M. & Grier, D. (1987). *Phys.Rev A* 35, pp. 4012
Rathgeber, S.; Monkenbusch, M.; Hedrick, J. L.; Trollsås, M. & Gast, A. P. (2006). *J. Chem. Phys.* 125, pp. 204908
Ratner, M.A. (1987), in 'Polymer Electrolyte Reviews I, Ed. J.R. MacCallum and C.A. Vincent, Elseiver Applied Science, London, 173-236
Rosso, M. (2007) *Electrochimica Acta* 53 250–256
Rucker, R. (1984), *The Fourth Dimension*, Houghton-Mifflin
Saffman P. G. & Taylor G., (1958). The penetration of a fluid into a medium of hele-shaw cell containing a more viscous liquid. *Proc. Soc. London, Ser A*, pp. 312-329.
Sawada, Y., Dougherty, A., Gollub, J.P. (1986): *Phys. Rev. Lett.* 56, pp. 1260.
Shibkov, A. A., Golovin Yu. I., Zheltov, M. A., Korolev A. A. & Vlasov A. A., (2001). Kinetics and morphology of nonequilibrium growth of ice in supercooled water, *Crystallography Reports*, 46, 3, pp. 496-502.
Shui, J.L., Jiang, G.S. Xie, S. & Chen, C.H. (2004) *Electrochimica Acta* 49 2209–2213
Sławinski, C., Sokołowska, Z. Walczak, R., Borówko, M. & Sokołowski, S. (2002). Fractal dimension of peat soils from adsorption and from water retention experiments, *Colloids and Surfaces, A: Physicochemical and Engineering Aspects* 208, pp. 289–301.
Smith Jr. T.G., Marks W.B., Lange G.D., Sheriff Jr. W.H. & Neale E.A. (1990). A fractal analysis of cell images. *J.Neurosci.Meth.* 27, pp. 173-180.
Stanley, H. E. , Buldyrev, S. V. , Goldberger, A. L. , Havlin S. , Mantegna, R. N. , Ossadnik, S. M., Peng, C. -K. , Sciortino, F. & Simons, M.(1994). *Fractals in Biology and Medicine: Lecture notes in Physics*. USA:Springer .
Suki, M.N.; Mohamed, N.S.; Hashim Ali, S.A. & Zainuddin, R. (2007). The role of image processing in measuring fractal dimension, *Malaysian Journal of Science*, pp. 23-33
Tordoff, G. M. Boddy, L. & Hefin Jones, T.(2007), Species-specific impacts of collembola grazing on fungal foraging ecology, *Soil Biology & Biochemistry*, pp. 1-9.
Vicsek, T. (1992) *Fractal Growth Phenomena*, second edition World Scientific Publishing
Villani, V. and Comenges, J. M. Z.(2000). Analysis of biomolecular chaos in aqueous solution. *Springer* 104, pp. 3-4.
West, B.J. & A.L. Goldberger.(1987). Physiology in fractal dimensions. *Am. Sci.* 75, pp. 354-365.
Wiens, J.A. (1989). Spatial scaling in ecology. *Funct. Ecol.* 3, pp. 385-397.
Witten Jr, T. A. & Sander, L. M. (1981). *Phys. Rev. Lett.* 47, pp. 1400.
Yadegari, S., Self-similarity, 16.03.2007, Available from http://www.crca.ucsd.edu/~syadegar/MasterThesis/node25.html
Zhang, G., Jin, L., Ma, Z., Zhai, X., Yang, M., Zheng, P., Wang, W. & Wegner, G. (2008). *J. Chem. Phys.* 129, pp. 224708